THE FUTURE
OF MAN

THE FUTURE
OF MAN

PIERRE TEILHARD DE CHARDIN

*Translated from the French
by Norman Denny*

IMAGE BOOKS
DOUBLEDAY
*New York London Toronto
Sydney Auckland*

AN IMAGE BOOK
PUBLISHED BY DOUBLEDAY
a division of Random House, Inc.

IMAGE, DOUBLEDAY, and the portrayal of a deer drinking from a stream
are registered trademarks of Random House, Inc.

Book design by Nicola Ferguson

Library of Congress Catalog Card Number: 64-20327

ISBN 978-0-385-51072-1

PRINTED IN THE UNITED STATES OF AMERICA

First Image Books Edition

May 2004

"The whole future of the Earth, as of religion, seems
to me to depend on the awakening of
our faith in the future."

PIERRE TEILHARD DE CHARDIN,
LETTERS TO MME GEORGES-MARIE HAARDT

This book consists of papers dealing with different aspects of a common theme, written by Pierre Teilhard de Chardin over a period of thirty years and here presented in their chronological order.

CONTENTS

TRANSLATOR'S NOTE

LIKE EVERY THINKER exploring new fields of thought, Père Teilhard de Chardin was faced by the problem of the limitations of language. He solved it, or got round it, in the way philosophers and scientists have always been obliged to do—by the use of neologisms and, at times, of elaborate, allusive formulations of words which make considerable demands on the reader if their full meaning and implications are to be grasped.

The difficulties confronting his translator need not be stressed. Limitations differ from language to language. There are things calling for laborious exposition in French, which can be said more clearly and simply in English; and the reverse is equally true. The problems have to be solved as best they can, often in an arbitrary way. I must cite one instance, which may otherwise puzzle the reader of this book. This is the word *Reflection*, which is also spelt *Reflexion* where the context seems to require it.

In Teilhard's philosophy, to which it is vital, the word represents two distinct things, which, however, are so intimately connected as to be in effect different aspects of the same thing. *Reflection* is the power of conscious thought which distinguishes

Man from all other living creatures (the animal that not only knows but *knows that it knows*). But the species Man also differs biologically, in Teilhard's view, from all other species (or phyla) in that, instead of spreading out fanwise, breaking into subspecies and falling eventually into stagnation, it coils inward upon itself and thus generates new (spiritual) energies and a new form of growth—a process of *Reflexion* which is part and parcel of the phenomenon of *Reflection*. I have varied the spelling according to which aspect appears more important in terms of the immediate argument.

The aim of this translation is twofold. First, and obviously, to convey Teilhard's meaning as clearly as possible. Second, and no less importantly, to catch the sound of his voice: to convey something of the nature of the man himself as it emerges from his writings, his warmth and humanity, his eager, wide-ranging, wonderfully lucid and penetrating mind, and above all, his passionate desire to impart what he had to say to everyone who will trouble to listen. Its success in achieving these aims can, at the best, be only relative. There never has been, nor ever will be, a "total" translation—anyway, of any sentence longer than half a dozen words.

I am most grateful to Mrs. Helen Suggett, who scrutinized my text with meticulous care, drew attention to many shortcomings and made many helpful suggestions.

Norman Denny

THE FUTURE
OF MAN

CHAPTER 1

A NOTE ON PROGRESS

e pur si muove

THE CONFLICT DATES from the day when one man, flying in the face of appearance, perceived that the forces of nature are no more unalterably fixed in their orbits than the stars themselves, but that their serene arrangement around us depicts the flow of a tremendous tide—the day on which a first voice rang out, crying to Mankind peacefully slumbering on the raft of Earth, "We are moving! We are going forward!" . . .

It is a pleasant and dramatic spectacle, that of Mankind divided to its very depths into two irrevocably opposed camps—one looking toward the horizon and proclaiming with all its new-found faith, "We are moving," and the other, without shifting its position, obstinately maintaining, "Nothing changes. We are not moving at all."

These latter, the "immobilists," though they

lack passion (immobility has never inspired anyone with enthusiasm!)[1], have commonsense on their side, habit of thought, inertia, pessimism and also, to some extent, morality and religion. Nothing, they argue, appears to have changed since man began to hand down the memory of the past, not the undulations of the earth, or the forms of life, or the genius of Man or even his goodness. Thus far practical experimentation has failed to modify the fundamental characteristics of even the most humble plant. Human suffering, vice and war, although they may momentarily abate, recur from age to age with an increasing virulence. Even the striving after progress contributes to the sum of evil: to effect change is to undermine the painfully established traditional order whereby the distress of living creatures was reduced to a minimum. What innovator has not retapped the springs of blood and tears? For the sake of human tranquillity, in the name of Fact, and in defense of the sacred Established Order, the immobilists forbid the earth to move. Nothing changes, they say, or can change. The raft must drift purposelessly on a shoreless sea.

But the other half of mankind, startled by the lookout's cry, has left the huddle where the rest of the crew sit with their heads together telling time-honored tales. Gazing out over the dark sea they study for themselves the lapping of waters along the hull of the craft that bears them, breathe the scents borne to them on the breeze, gaze at the shadows cast from pole to pole by a changeless eternity. And for these all things, while remaining separately the same—the ripple of water, the scent of the air, the lights in the sky—become linked together and acquire a new sense: the fixed and random Universe is seen to move.

[1] For the *status quo* of life as it exists: the "immobility" of the Christian, or of the Stoic, may arouse fervor because it is a *withdrawal*, that is to say an individual *anticipation* (more or less fictitious) of *consummated* progress.

No one in the world who has seen this vision can be restrained from guarding and proclaiming it. *To testify to my faith in it*, and to show reasons, *is my purpose here*.

———— ❋ ————

IT IS CLEAR in the first place that the world in its present state is the outcome of movement. Whether we consider the position of the rocky layers enveloping the Earth, the arrangement of the forms of life that inhabit it, the variety of civilizations to which it has given birth, or the structure of languages spoken upon it, we are forced to the same conclusion: that everything is the sum of the past and that nothing is comprehensible except through its history. "Nature" is the equivalent of "becoming," self-creation: this is the view to which experience irresistibly leads us. What can it mean except that the Universe must, at least at some stage, have been in movement; that it has been malleable, acquiring by degrees, not only in their accidental details but in their very essence, the perfections which now adorn it? There is nothing, not even the human soul, the highest spiritual manifestation we know of, that does not come within this universal law. The soul, too, has its clearly defined place in the slow ascent of living creatures toward consciousness, and must therefore in one way or another have grown out of the general mobility of things. Those who look reality in the face cannot fail to perceive this progressive genesis of the Universe, and with a clarity which leaves no room for doubt. Whatever the other side may say, clinging to their imaginary world, the Cosmos did once move, the whole of it, not only locally but in its very being. This is undeniable and we shall not discuss it further. But *is it still moving?* Here we have the real question, the living, burning question of Evolution.

------------------------------- ☀ -------------------------------

IT IS THE fundamental paradox of Nature as we see it now that its universal plasticity seems suddenly to have hardened. Like an ocean-wave caught in a snapshot, or a torrent of lava stiffened by cooling, the mountains and living things of the earth wear the aspect, to those who study them, of a powerful momentum that has become petrified. Nature seen at a distance appears to be malleable and in motion; but seek to lay hands on it, to deflect by force even the least of Life's directions, and you will encounter nothing but absolute rigidity, an unshakably stubborn refusal to depart from the preordained path.

But let us note that this present rigidity of Nature does not, as some people believe, in any way lessen the certainty of its past mobility. What we regard as the fixity of present organisms may be simply a state of very slow movement, or of rest between spells of movement. It is true that we have not yet succeeded in shaping life to our requirements in the laboratory; but who has shaped or witnessed the shaping of a geological stratum? The rock which we seek to compress crumbles because we work too fast or with oversmall fragments. Calcareous matter, if it is to be made malleable, needs to be embedded in a vast mass, and perhaps its reshaping is a process of immense slowness. If we have not seen the upward thrust of mountain ranges it is because their rise was accomplished either in widely spaced jerks or with so slow a rhythm that since the coming of Man nothing of the kind has happened, or at least nothing that has been perceptible to us. Why should not Life, too, be mobile only in great masses, or through the slow action of time, or in brief stages? Who can positively affirm that at this moment, although we perceive nothing, new forms are not taking shape in the contours of the earth and of Life? . . .

The plasticity of Nature in the past is an undeniable fact; its present rigidity is less capable of scientific proof. If we had to choose between transformism and fixism, that is to say between two absolutes—everything incestantly in motion, or everything for ever immovable—we should be bound to choose the first. But there is a third possible hypothesis, namely that everything was at one time fluid but is now *irrevocably fixed*. It is this third alternative that I wish to examine and dismiss.

———— ⁎ ————

THE HYPOTHESIS OF a definitive halt in terrestrial evolution is, to my mind, suggested less by the apparently unchanging nature of present forms than by a certain general aspect of the world coinciding with this appearance of cessation. It is most striking that the morphological change of living creatures seems to have slowed down at the precise moment when Thought appeared on earth. If we relate this coincidence to the fact that the only general line taken by biological evolution has been in the direction of the largest brain—broadly speaking, of the highest state of consciousness—we are compelled to wonder whether the true fundamental impulse underlying the growth of animal forces has not been the "need" to know and to think; and whether, when this overriding impulse eventually found its outlet in the human species, the effect was not to produce an abrupt diminution of "vital pressure" in the other branches of the Tree of Life. This would explain the fact that "evolving Life," from the end of the Tertiary era, has been confined to the little group of higher primates. We know of many forms that have disappeared since the Oligocene, but of no genuinely new species other than the anthropoids. This again may be explained by the extreme brevity of the Miocene as compared with other geological periods. But does it not lead us to surmise that the "phyla" possessing higher psychic attributes have absorbed all the forces at Life's disposal?

If we are to find a definitive answer to the question of the entitative progress of the Universe we must do so by adopting the least favourable position—that is to say, by envisaging a world whose evolutionary capacity is *concentrated upon* and *confined to the human soul.* The question of whether the Universe is still developing then becomes a matter of deciding whether the *human spirit* is still in process of evolution. To this I reply unhesitatingly, "Yes, it is." The nature of Man is in the full flood of entitative change. But to grasp this it is necessary (a) not to overlook the *biological* (morphogenic) value of moral action, and (b) to accept the organic nature of *interindividual* relationships. We shall then see that a vast evolutionary process is in ceaseless operation around us, but that it is situated within the *sphere of consciousness* (and collective consciousness).

WHAT IS THE difference between ourselves, citizens of the twentieth century, and the earliest human beings whose soul is not entirely hidden from us? In what respects may we consider ourselves their superiors and more advanced than they?

Organically speaking, the faculties of those remote forebears were probably the equal of our own. By the middle of the last Ice Age, at the latest, some human groups had attained to the expression of aesthetic powers calling for intelligence and sensibility developed to a point which we have not surpassed. To all appearance the ultimate perfection of the human *element* was achieved many thousands of years ago, which is to say that the individual instru-

ment of thought and action may be considered to have been final-
ized. But there is fortunately another dimension in which variation
is still possible, and in which we continue to evolve.

The great superiority over Primitive Man which we have
acquired and which will be enhanced by our descendants in a
degree perhaps undreamed-of by ourselves, is in the realm of self-
knowledge: in our growing capacity to situate ourselves in space and
time, to the point of becoming conscious of our place and responsibility in relation to the Universe.

Surmounting in turn the illusions of terrestrial flatness, immobility and autocentricity, we have taken the unhopeful surface of
the earth and "rolled it like a little ball"; we have set it on a course
among the stars; we have grasped the fact that it is no more than a
grain of cosmic dust; and we have discovered that a process without limit has brought into being the realms of substance and
essence. Our fathers supposed themselves to go back no further
than yesterday, each man containing within himself the ultimate
value of his existence. They held themselves to be confined within
the limits of their years on earth and their corporeal frame. We
have blown asunder this narrow compass and those beliefs. At
once humbled and ennobled by our discoveries, we are gradually
coming to see ourselves as a part of vast and continuing processes;
as though awakening from a dream, we are beginning to realize
that our nobility consists in serving, like intelligent atoms, the work
proceeding in the Universe. We have discovered that there is a
Whole, of which we are the elements. We have found the world in
our own souls.

What does this conquest signify? Does it merely denote the establishment, in worldly terms, of an idealized system of logical, extrinsic relationships? Is it no more than an intellectual luxury, as is
commonly supposed—the mere satisfaction of curiosity? No. The

consciousness which we are gradually acquiring of our physical relationship with all parts of the Universe represents a genuine en-
larging of our separate personalities. It is truly a progressive
realization of the universality of the things surrounding each of us.
And it means that in the domain external to our flesh our *real and
whole body is continuing to take shape.*

That is in no way a "sentimental" affirmation.

The proof that the growing coextension of our soul and the
world, through the consciousness of our relationship with all things,
is not simply a matter of logic or idealization, but is part of an organic process, the natural outcome of the impulse which caused
the germination of life and the growth of the brain—the proof is
that it expresses itself in a *specific evolution of the moral value of our actions* (that is to say, by the modification of what is most living within
us).

No doubt it is true that the scope of individual human action,
as commonly envisaged in the abstract theory of moral and meri-
torious acts, is not greatly enhanced by the growth of human
knowledge. Inasmuch as the willpower of contemporary man is
not in itself more vigorous or unswerving than that of a Plato or
an Augustine, and individual moral perfection is still to be measured by steadfastness in pursuance of the known good (and therefore relative) we cannot claim as *individuals* to be more moral or
saintly than our fathers.

Yet this must be said, to our own honor and that of those who
have toiled to make us what we are: that between the behavior of
men in the first century A.D. and our own, the difference is as great,
or greater, than that between the behavior of a fifteen-year-old boy
and a man of forty. Why is this so? Because, owing to the progress
of science and of thought, our actions today, whether for good or
ill, proceed from an incomparably higher point of departure than

those of the men who paved the way for us toward enlightenment. When Plato acted it was probably in the belief that his freedom to act could only affect a small fragment of the world, narrowly circumscribed in space and time; but the man of today acts in the knowledge that the choice he makes will have its repercussions through countless centuries and upon countless human beings. *He feels in himself the responsibilities and the power of an entire Universe.* Progress has not caused the *action of Man* (Man himself) to change in each separate individual; but because of it the *action of human nature* (Mankind) has acquired, in every thinking man, a fullness that is wholly new. Moreover, how are we to compare or contrast our acts with those of Plato or Augustine? All such acts are linked, and Plato and Augustine are still expressing, through me, the whole extent of their personalities. There is a kind of human action that gradually matures through a multitude of human acts. The human monad has long been constituted. What is now proceeding is the animation (assimilation) of the Universe by that monad; that is to say, the realization of a *consummated human Thought.*

There are philosophers who, accepting this progressive animation of the concrete by the power of thought, of Matter by Spirit, seek to build upon it the hope of a terrestrial liberation, as though the soul, become mistress of all determinisms and inertias, may someday be capable of overcoming harsh probability and vanquishing suffering and evil here on earth. Alas, it is a forlorn hope; for it seems certain that any outward upheaval or internal renovation which might suffice to transform the Universe as it is could only be a kind of death—death of the individual, death of the race, death of the Cosmos. A more realistic and more Christian view shows us Earth evolving toward a state in which Man, having come into the full possession of his sphere of action, his strength, his maturity and his unity, will at last have become an adult being;

and having reached this apogee of his responsibility and freedom, holding in his hands all his future and all his past, will make the choice between arrogant autonomy and loving excentration.

This will be the final choice: whether a world is to revolt or to adore.[2] And then, on an act which will summarize the toil of centuries, on this act (finally and for the first time completely human) justice will set its seal and all things be renewed.

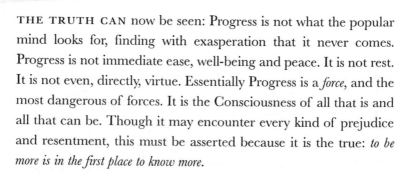

THE TRUTH CAN now be seen: Progress is not what the popular mind looks for, finding with exasperation that it never comes. Progress is not immediate ease, well-being and peace. It is not rest. It is not even, directly, virtue. Essentially Progress is a *force*, and the most dangerous of forces. It is the Consciousness of all that is and all that can be. Though it may encounter every kind of prejudice and resentment, this must be asserted because it is the true: *to be more is in the first place to know more.*

Hence the mysterious attraction which, regardless of all setbacks and *a priori* condemnations, has drawn men irresistibly toward science as to the source of Life. Stronger than every obstacle and counterargument is the instinct which tells us that, to be faithful to Life, we must *know*; we must know more and still more; we must tirelessly and unceasingly search for Something, we know not what, which will appear in the end to those who have penetrated to the very heart of reality.

I maintain that it is possible, by following this road, to find substantial reasons for belief in Progress.

---

[2] My purpose is not to show that a *necessary* or *infallible* line of progress exists, but simply to establish that, *for Mankind as a whole*, a way of progress is offered and awaits us, analogous to that which the individual cannot reject without falling into sin and damnation.

The world of human thought today presents a very remarkable spectacle, if we choose to take note of it. Joined in an inexplicable unifying movement men who are utterly opposed in education and in faith find themselves brought together, intermingled, in their common passion for a double truth; namely, that there exists a physical Unity of beings, and that they themselves are living and active parts of it. It is as though a new and formidable mountain chain had arisen in the landscape of the soul, causing ancient categories to be reshuffled and uniting higgledy-piggledy on every slope the friends and enemies of yesterday: on one side the inflexible and sterile vision of a Universe composed of unalterable, juxtaposed parts, and on the other side the ardor, the faith, the contagion of a living truth emerging from all action and exercise of will. Here we have a group of men joined simply by the weight of the past and their resolve to defend it; there a gathering of neophytes confident of their truth and strong in their mutual understanding, which they feel to be final and complete.

There seem to be only two kinds of mind left; and—it is a disturbing thought—all natural mystical power and all human religious impulse seem to be concentrated on one side. What does this mean?

There are people who will claim that it is no more than a mode, a momentary ripple of the spirit—at the most the passing exaggeration of a force that has always contributed to the balance of human thinking. But I believe we must look for something more. This impulse which in our time is so irresistibly attracting all open minds toward a philosophy that comprises at once a theoretical system, a rule of action, a religion and a presentiment, heralds and denotes, in my view, the effective, physical fulfillment of all living beings.

We have said that progress is designed to enable considered action to proceed from the willpower of mankind, a wholly human exercise of choice. But this natural conclusion of the vital effort, as we can now see, is not to be regarded as something consummated separately in the secret heart of each monad. If we are to perceive and measure the extent of Progress we must look resolutely beyond the individual viewpoint. It is Mankind as a whole, collective humanity, which is called upon to perform the definitive act whereby the total force of terrestrial evolution will be released and flourish; an act in which the full consciousness of each individual man will be sustained by that of every other man, not only the living but the dead. And so it follows that the *opus humanum* laboriously and gradually achieved within us by the growth of knowledge and in the face of evil, is something quite other than an act of higher morality: it is a living organism. We cannot distinctly view its progress because the organism encloses us, and to know a thing synthetically one has to be able to see it as a whole. Yet is there any part of ourselves which does not glow and responsively vibrate with the measure of our growth?

We need only to look about us at the multitude of disjointed forces neutralizing each other and losing themselves in the confusion of human society—the huge realities (broad currents of love or hatred animating people and classes) which represent *consciousness in potency* but have not yet found a consciousness sufficiently vast to encompass them all. We need only recall those moments in time of war when, wrested out of ourselves by the force of a collective passion, we have a sense of rising to a higher level of human existence. All these spiritual reserves, guessed at and faintly apprehended, what are they but the sure evidence that creation is still on the move, but that we are not yet capable of expressing all the natural grandeur of the human mission?

Vistas such as these, I know, do not appear to come within the Christian perspective; and because of this most of those who point to them and welcome them seem, at least by implication, to be heralding the appearance of a religion destined to supplant all earlier creeds. But how does it all arise—the challenge on the one hand, and the mistrust on the other—except out of the fact that neither we nor our adversaries have sufficiently measured the powers of growth with which Christ endowed his Church?

For my own part I accept the reality of the movement which tends to segregate, within the bosom of Mankind, a congregation of the faithful dedicated to the great task, "Advance in unity!" Moreover, I believe in its truth; I consider the fact that it contains in its ranks a great number of sinners, of "the maimed, and the halt, and the blind," to be evidence of this truth. But this does not cause me to believe that the eager multitude crying out today for guidance is in search of any Shepherd other than He who has already brought it bread.

Christ, as we know, fulfills Himself gradually,[3] through the ages in the sum of our individual endeavors. Why should we treat this fulfillment as though it possessed none but a metaphorical significance, confining it entirely within the abstract domain of purely supernatural action? Without the process of biological evolution, which produced the human brain, there would be no sanctified souls; and similarly, without the evolution of collective thought, through which alone the plenitude of human consciousness can be attained on earth, how can there be a consummated Christ? In other words, without the constant striving of every human cell to unite with all the others, would the Parousia be physically possible? I doubt it.

That is why I believe that this coming together, from all four corners of the intellectual world, of a great mass of naturally reli-

---

[3] In his Mystical Body: cf. the last paragraph of Cosmic Life, p. 307. (Ed.).

gious spirits, does not portend the building of a new temple on the
ruins of all others but the laying of new foundations to which the
old Church is gradually being moved.

Little by little the idea is coming to light in Christian consciousness that the "phylogenesis" of the whole man, and not
merely the "ontogenesis" of his moral virtues, is hallowed, in the
sense that the charity of the believer may more resemble an impulse of constructive energy and his self-detachment be more in
the nature of a positive effort.

In response to the cry of a world trembling with the desire for
unity, and already equipped, through the workings of material
progress, with the external links of this unity, Christ is already revealing himself, in the depths of men's hearts, as the Shepherd (the
Animator) of the Universe. We may indeed believe that the time is
approaching when many men, old and new believers, having understood that from the depths of Matter to the highest peak of the
Spirit there is only *one evolution*, will seek the fullness of their
strength and their peace in the assured certainty that the whole of
the world's industrial, aesthetic, scientific and moral endeavor
serves physically to complete the Body of Christ, whose charity animates and re-creates all things.

Fulfilling the profound need for unity which pervades the
world, and crowning it with renewed faith in Christ the Physical
Center of Creation; finding in this need the natural energy required for the renewal of the world's life; thus do I see the New
Jerusalem, descending from Heaven and rising from the Earth.

He who speaks these words before the Tribunal of the Elders will
be laughed at and dismissed as a dreamer.

"Nothing moves," a first Sage will say. "The eye of common sense sees it and science confirms it."

"Philosophy shows that nothing can move," says a second.

"Religion forbids it—nothing must move," says a third.

Disregarding this triple verdict the Seer leaves the public place and returns to the firm, deep bosom of Nature. Gazing into the depths of the immense complex of which he is a part, whose roots extend far below him to be lost in the obscurity of the past, he again fortifies his spirit with the contemplation and the feeling of a universal, stubborn movement depicted in the successive layers of dead matter and the present spread of the living. Gazing upward, toward the space held in readiness for new creation, he dedicates himself body and soul, with faith reaffirmed, to a Progress which will bear with it or else sweep away all those who will not hear. His whole being seized with religious fervor he looks toward a Christ already risen but still unimaginably great, invoking, in the supreme homage of faith and adoration, *"Deo Ignoto."*

HITHERTO UNPUBLISHED. PARIS, AUGUST 10, 1920.

CHAPTER 2

SOCIAL HEREDITY
AND PROGRESS

Notes on the Humano-Christian Value
of Education

1. Education and Life

TO THE EYE of Physical Science, one of the most remarkable characteristics of Life is its "additive" quality. Life propagates itself by ceaselessly adding to itself what it successively acquires—like a memory, as has often been said. Every living being passes on to his successor the being he himself inherited, not merely diversified but *accentuated* in a given direction, according to the line to which he belongs. And all the lines, whatever their nature, seem in varying degrees and each after its own formula to move a greater or lesser distance in the general direction of greater spontaneity and consciousness. Something passes, something grows, through the long chain of living creatures. This is the great fact, or the great law, whose discovery has transformed our vision of the Universe during nearly two centuries.

At what levels and by what mechanisms does this predetermined additivity of characteristics show itself in the living being?

An essential part of the phenomenon must take place at the moment of reproduction. The wave of life in its substance and with its particular characteristics is of necessity communicated to the child in and through the fertilized cell, the issue of the parents. Fundamentally, biological evolution can only be an effect of germinal transmission. That is why the science of Life concentrates more and more upon the study of cellular heredity.

But a difficulty arises. As we have said, it appears to be the case that every zoological chain observed over a sufficiently long period can be seen to modify itself in a given direction (shape of limbs or teeth, relative development of the brain, etc.), so that certain specific characteristics are found to have increased throughout the part of the chain under observation. Something has undoubtedly been gained, yet it would seem that none of the elements in the chain, taken separately, has *actively* contributed to this gain. Although it was accepted without discussion in the early days of transformism, the question of the germinal transmission to the children of characteristics *acquired* by the parents has become one of those most hotly disputed among geneticists. No irrefutable evidence of any such transmission has yet been found, and there are now many biologists who flatly deny that it takes place. But this amounts to saying that the individual links in a biological chain passively transmit a germ evolved in themselves, without in any way affecting it by their own activities: the bodies (the "somata") grow out of this "germen" which is inexplicably endowed with its own power of evolutionary development; they are its dependents but incapable of modifying it. It is a highly improbable hypothesis, having the grave disadvantage that it deprives the individual of all responsibility in the development of the race or the particular branch of which he is a part.

For the purpose of examining the additive mechanism of Life in its vital, active form I propose to look in a direction which the theorists of heredity seem to have ignored. No complete light has yet been thrown on the secret processes taking place in the microscopic recesses of the cell. Let us turn instead to a phenomenon that we can clearly see because it is on our own scale, and note what happens in the field of *education*.

Education. The transmission by example of an improvement, an action, and its reproduction by imitation. We are curiously inclined to minimize the significance and the import of this function in the development of Life, for a variety of reasons. Education is so widespread a phenomenon, so clearly visible, humble and commonplace, that there seems to be no reason to look for any mystery in it. Moreover, it appears to be so exclusively associated with the human condition that it is hard to attribute to it any universal biological value. Finally, it is such a fragile and superficial structure, shedding a haphazard light on our lives, maintaining and propagating itself by grace of circumstances that are in themselves precarious and changeable: how can we compare it to those deep, underlying determinisms which impose an ineluctable course upon the advance of Life?

These various arguments or appearances, confusedly perceived and accepted, undoubtedly divert our attention from the "educational factor," causing us to dismiss it as an "epiphenomenon" unworthy of the attention of the natural scientist and the physicist. Yet there is not one of them that cannot be revised or reversed to sustain a precisely opposite thesis.

Education is infinitely commonplace. . . . But what could be more ordinary than the three dimensions in space, the fall of a body, the propagation of light, the growth of a plant? What does the fundamental progress of science consist in, except the discovery of the organic, structural value of what is most general and everyday in our experience?

Education is a specifically human phenomenon. . . . No doubt, where it is a question of reasoned education! But we have only to observe the animal world with minds more open to the ideas of birth and evolution to perceive, in this as in every case, that the "human" could not exist if it did not contain, transfigured in terms of mind, a property common to all animals, of which the beginnings are to be detected as they vanish into the past stretching behind us. The dog, the cat or the bird train their young in countless ways to hunt, to fly or to build a nest. The monkey does much more. And how are we to explain the remarkable behavior patterns of the beaver, or of insects, except as the outcome of accumulated and transmitted experiences and discoveries? Such phenomena become apparent to us only where the creature under study has attained a sufficient degree of spontaneity, still more if it lives in a group. But what more is needed to persuade us that, at least for practical purposes, education is a universal biological function, coextensive with the totality of the living world?

We may be tempted to add, nevertheless, that education is an extrinsic mechanism, superposed at one remove on the transmission of life. But Bergson has pointed out the arbitrary nature of the dividing line drawn by common sense between the zone of "organic" determinisms and that of "spontaneity" in the course of embryogenesis. When the chick pecks its way out of the egg, is it the "germen" or the "soma" that guides its beak? The same insidious question, perfectly justified in the case of "ontogenesis," arises again and no less embarrassingly when it comes to the generative process itself. At what point does the mother cease to engender her child? Is it when she first feeds it, after giving it birth? Or is it when, having weaned it, she teaches it to know and hunt its prey? In fact, and although it operates successively on two different levels (that of the purely organic functioning of the mother, and that of her conscious action applied to another consciousness), what takes place is one and the same process pursued externally from one end of the chain to the other. This leads to that; and this is probably capable of acting upon that. We have spoken of the biologists who reject the germinal transmission of acquired characteristics. Have they considered the case of the countless insects which, dying without having known their progeny, nevertheless transmit behavior patterns to descendants which they never see? If these patterns, as it seems we must assume, were discovered by spontaneous experiment at a time when, owing to a different arrangement of the seasons, or of lives or metamorphoses, the parents knew and reared their young, then this in fact means that the results of education finally entered into the germ itself, endowing it with attributes as physically predetermined as size or color or any other of the inherited characteristics of the species or breed.

So we reach the following conclusion, which seems to me valid. Far from being an artificial, accidental, or accessory phenomenon in its relation to living creatures, education is nothing less than an essential and natural form of biological additivity. In it we can perhaps catch a glimpse, still in the marginal, conscious state, of individual, germinal heredity in process of formation: as though organic mutation at this stage took the form of a psychic invention contrived by the parents and transmitted by them. And also—this is the least that can be said—we see heredity pass through education beyond the individual to enter into its collective phase and become social.

The first and most evident outcome of this view of the matter is the singular extent to which it coordinates and unifies such ideas as we have been able to arrive at on the subject of life in general. But it has another advantage which I particularly wish to dwell upon. It sheds a new light on the importance and dignity of everything that affects the education of Mankind.

2. *Education and Mankind*

LIFE HAD ATTAINED through Man the highest degree of inventive choice in the individual and socialization in the community. For this double reason the phenomenon of education as it affects Man possesses a greater amplitude and clarity than in any other context and calls for more exhaustive study.

Breathing the atmosphere of human education as we do from the moment of our birth, we have little inclination or time to consider what it represents, either on its own account or in relation to ourselves. Yet if we pause to look we can find much to make us marvel. The following experiment is worth making. Let us imagine ourselves to be divested of everything that we owe to life in human society. To begin with we must eliminate all the latest modes of communication (surface, air, radio) devised by science. But we need to go much further than this. We must cut ourselves off from industry and agriculture; we must forget our history; we must assume that even language does not exist. In short, we must get as close as we can to that almost inconceivable state in which our consciousness, divorced from all human association, stands naked in face of the Universe. What is then left of our essential self? Have we in our mind's eye merely shed the garments from our body, or a part of our very soul? . . . Now let us reverse the process, reclothing ourselves piece by piece with those layers of education which we imagined we tried to cast aside. But in doing so let us seek, however confusedly, to re-create what we can of their history. What immeasurable toil went into the weaving of each garment, what endless time, what trial and error, what a countless multitude of hands! Thinking of this we may be disposed to say, "It is all an accessory and very fragile. A single catastrophe, bringing the whole of that secular edifice down in ruins, could cause Man to revert to his earliest state, when Thought was first born on earth." Yet how can we fail to perceive in that patient and continuous amassing of human acquirements the methods and therefore the very stamp of Life itself—Life which is *irreversible*, its inevitability born of the improbable, its consistence of fragility.

Let us rather accept the fact: Mankind, as we find it in its present state and present functioning, is organically inseparable from that which has been slowly added to it, and which is propagated through education. This "additive zone," gradually created and transmitted by collective experience, is for each of us a sort of matrix, as real in its own way as our mother's womb. It is a true racial memory, upon which our individual memories draw and through which they complete themselves. Applied to the particular and singular instance of the human species, the idea that education is not merely a "subphenomenon," but an integral part of biological heredity, derives unquestionable verification from the very coherence which it brings to the whole landscape, and the relief into which it throws it.

But we must logically go a step further. The additivity of organic life, as science now tells us, is something quite different from the superposition of characteristics added to one another like the layers forming a sedimentary deposit. Life does not merely "snowball"; it behaves more like a tree, which acquires successive rings according to the particular fashion of its growth, in a *directed* manner. To accept that education is one of the factors, or better, one of the forms of the process which we denote by the very generalized and rather vague term evolution, is therefore to imply that the sum of knowledge and acquirement retained and transmitted by education from one generation to the next constitutes a natural sequence of which the direction may be observed.

And this is precisely what happens.

It may seem difficult, at first glance, to distinguish any kind of order in the jumble of experiments, organizations and theories whose incessantly growing mass forms the baggage-train of the human caravan. Purely quantitative progress, the sceptics tell us. But if we stand back a little, and look at the phenomenon as a whole, we can see that all is not confusion. For it then becomes apparent that this accumulation of features, bewildering at close quarters, does in fact outline a face: the face of Mankind gradually acquiring the knowledge of its birth, its history, its natural environment, its external powers, and the secrets of its soul.

"That which takes place in all of us when, as we grow up, we become aware of our family past, our present responsibilities, our ambitions and our loves, is nothing but the brief recapitulation of a far vaster and slower process through which the whole *human race* must pass in its growth from infancy to maturity." . . . We have heard this said many times; but have we pondered it to the point of realizing the full intensity and extent of its truth? It denotes the reality of a growth of Mankind through and above the growth of individual men. . . . No doubt it is true, if we judge by the written word, that we cannot claim to be more intelligent than our fathers. But it is undeniable that, thanks to their accumulated efforts, we have a better understanding than they could possess of the dimensions, the demands, potentialities and hopes; above all of the profound unity of the world within and around us. In the passage of time a state of collective human consciousness has been progressively evolved which is inherited by each succeeding generation of conscious individuals, and to which each generation adds something. Sustained, certainly, by the individual person, but at the same time embracing and shaping the successive multitude of individuals, a sort of generalized human personality is visibly in process of formation upon the earth. It seems that where Man is concerned the specific function of education is to ensure the continued development of this personality by transmitting it to the endlessly changing mass: in other words, to extend and ensure in collective mankind a consciousness which may already have reached its limit in the individual. Its fulfillment, in the case of man, of this specific function is the final proof of the biological nature and value of education, extending to the things of the spirit.

3. Education and Christianity

SINCE THESE LINES are intended for Christian teachers I must now transpose the ideas I have outlined to the dimensions of the Christian supernatural. How do these ideas work out, and to what extent can they be fully developed in this new domain of Creation?

By definition and in essence Christianity is the religion of the Incarnation: God uniting Himself with the world which He created, to unify it and in some sort incorporate it in Himself. To the worshipper of Christ this act expresses the history of the universe.

But how does it operate, this gradual conquest and assimilation of Earth by Heaven? In the first place quantitatively, by the addition to the Mystical Body of an increasing multitude of human souls, "until the number shall be complete." But also qualitatively, by the steady growth, within the bosom of the Church, of a certain Christological perspective. Through the living *tradition* of a faith and a mystique the Christian organism diffuses or expresses in itself an ever more awakened sense of Christ present and active in the fulfillments of the world. We cannot continue to love Christ without discovering Him more and more. The maturing of a collective consciousness accompanied by numerical expansion: these are two aspects inseparably linked in the historical unfolding of the Incarnation.

And so in Christianity we again come upon that mysterious law of additivity and social heredity which in every field governs the processes of Life; while at the same time in this new domain the fundamental role of education is again manifest, as the human instrument of divine instruction. But a new and fascinating prospect also emerges. As we have said, human endeavor, viewed in its "natural" aspect, is tending toward some sort of collective personality, through which the individual will acquire in some degree the consciousness of Mankind as a whole. Viewed on the other hand in its "supernatural" aspect this endeavor expresses itself and culminates in a sort of participation in the divine life, whereby each individual will find, by conscious union with a Supreme Personal Being, the consummation of his own personality. Is it conceivable that two cases bearing so much resemblance can be wholly divorced from one another? Or are these two trends of collective consciousness, one toward Christ, the other toward Mankind, simply related phases, on different levels, of the same event?

To postulate the truth of the second alternative—that is to say, to accept that in terms of the divine purpose the two impulses are one—is to define in its essentials, and in all its splendor, the attitude of *Christian humanism.*

To the Christian humanist—faithful in this to the most sure theology of the Incarnation—there is neither separation nor discordance, but coherent subordination, between the genesis of Mankind in the World and the genesis of Christ in Mankind through His Church. The two processes are inevitably linked in their structure, the second requiring the first as the matter upon which it descends in order to superanimate it. This view entirely respects the progressive, and experientially known, concentration of human thought in an increasingly acute consciousness of its unitary destiny. But instead of the vague center of convergence envisaged as the ultimate end of this process of evolution, the personal and defined reality of the Word Incarnate, in which everything acquires consistence, appears and takes its place.

Life for Man. Man for Christ. Christ for God.

And to ensure the psychic continuity, at every phase, of this vast develop-
ment embracing myriads of elements strewn throughout the immensity of time,
there is a single mechanism—education.

All the lines join together, complete themselves and merge.
Everything becomes one whole.

Which brings us to this final summing up, wherein is revealed
the gravity and unity, but also the complexity, of the seemingly
humble task of the Christian educator:

a It is primarily through education that the hereditary biological process, which from the beginning has caused the world to
rise to ever higher zones of consciousness, is furthered in a reflective form and in its social dimensions. The educator, as an instrument of Creation, should derive respect and ardor for his efforts
from a profound, communicative sense of the developments already achieved or awaited by Nature. Every lesson he gives should
express love for, and cause to be loved, all that is most irresistible
and definitive in the conquests of Life.

b It is through education, by the progressive spread of common viewpoints and attitudes, that the slow convergence of minds
and hearts is proceeding, without which there seems to be no outlet ahead of us for the impulse of Life. Directly charged with the
task of achieving this unanimity of mankind, the educator,
whether his subject be literature, history, science or philosophy,
must constantly live with it and consciously strive for its realization.
A passionate faith in the purpose and splendor of human aspirations must be the flame that illumines his teaching.

c Finally, it is through the medium of education that there
ensues, directly and indirectly, the gradual incorporation of the

World in the Word Incarnate: indirectly, in the degree in which the heart of a collective Mankind increasingly turned inward upon itself is made ready for this high transformation; directly, to the extent that the tide of Grace historically released by Jesus Christ is propagated only by being borne on a living tradition. But the teacher who seeks to be wholly effective in transmitting these two influences, the humanizing and the Divine, must be as it were overwhelmed by the evidence of their inseparable, structural relation. To have experienced and understood, in order to teach others to experience and understand, that all human enrichment is but dross except inasmuch as it becomes the most precious and incorruptible of all things by adding itself to an immortal center of love: such is the supreme knowledge and the ultimate lesson to be imparted by the Christian educator.

These three linked propositions complete a logical structure whose perfect harmony proclaims its truth.

In the present day human education is spreading its net over the earth on an unprecedented scale and by means of unprecedented methods of expression and diffusion. Never have there been so many libraries, periodicals, schools, universities, laboratories—or pupils! And it is remarkable that in this magnificent whole, proportionate in scale to the new age which we are entering, there is no institution, other than Christianity, that seems capable of endowing the immense body of things taught with a true soul. Because he alone has the power to invest human endeavor and enrichment with positive aspirations and a positive objective, the Christian teacher alone is in a position to fulfill, both in the consciousness he employs and the consciousness he transmits, the total function of the educator.

1938. *ÉTUDES*, APRIL 1945.

CHAPTER 3

THE GRAND OPTION

1. On the Threshold of Human Socialization

JUST AS ASTRONOMY, by the comparative study of heavenly bodies, has been able to detect the existence and determine the phases of a life-history of the stars, so the science of biology, by the comparative study of living forms, has been able to determine the successive stages through which animal and vegetable groups pass in the course of their evolution. No natural scientist doubts any longer that different species appear, grow, age and die.

In this sense it is evident that Mankind of its nature behaves like a species, and is therefore subject as a whole, as in the case of the individual, to a definite cycle of development. To every thinking man this poses a problem of obvious importance for the ordering and orientation of our collective life. What is the precise point reached *at this moment* by the human race in the ineluctable curve of growth which is described by every zoological species in the course of its existence? In other words, what phase of its "phyletic" development may we consider Mankind to have attained at the present time, in comparison with the other branches surrounding us on the tree of life?

This overwhelmingly important question is one to which I think we can find a reply provided we take into account a phenomenon familiar to all biologists, but of which the significance in terms of "phase" or "stage" has not been sufficiently recognized or made use of: I mean that of *association* or, better still, *social organization*. No sooner is it constituted by the grouping together of elementary particles, than the living element, whatever its degree of internal complexity, begins to reproduce itself. But the process does not end there. When it exists in sufficient numbers the separate element tends to link up with others of its kind so as to form with them a more or less differentiated organic whole. In this fashion the higher plants and the metazoa evolved out of isolated cells, the corals out of fixed or drifting polyps, the termitary out of free neuroptera and the ant hill and the bee colony out of independent hymenoptera. A similar impulse of group formation seems to have become operative along each zoological branch, but at very different ages of the earth; so far as we are able to judge, the phenomenon has occurred in each case at a predetermined age of the species under review. In the case of the oldest groups the mechanism of their formation can only be conjectured; but with more recent groupings the stages of the process may still be discerned in their present natural state. We know of unattached bees and wasps, and of others forming small and loosely ordered communities; and by way of a varying series of intermediate states we arrive at the bee colony, which is almost as organically centered on its queen as is the termitary. *In short, everything happens as though, in the course of its phyletic existence, every living form achieved (with more or less success) what may be called a period, or even a point, of socialization.*

This being so, let us look at the human species and see if we can fit it into the scheme. Because we are a part of it, because the rhythm of its growth is infinitely slow in comparison with our own, and because its grandeur overwhelms us, Mankind, in its total evolution, escapes our intuitive grasp. But may we not see reflected in the life around us things that we cannot see directly in ourselves? Let us study ourselves in the mirror of other living forms. What do we see?

Prehistory teaches us that in the beginning Man must have lived in small, autonomous groups; after which links were established, first between families and then between tribes. These associations became more elaborate as time went on. In the phase of the "neolithic revolution" they hardened and became fixed on a territorial basis. For thousands of years this principle remained essentially unchanged; it was the land, despite all social readjustments, which remained the symbol and the safeguard of individual liberty in its earliest form. But now a further transformation is taking place; it has been going on irresistibly for a century under our very eyes. In the totalitarian political systems, of which time will correct the excesses but will also, no doubt, accentuate the underlying tendencies or intuitions, the citizen finds his center of gravity gradually transferred to, or at least aligned with, that of the national or ethnic group to which he belongs. This is not a return to primitive and undifferentiated cultural forms, but the emergence of a defined social system in which a purposeful organization orders the masses and tends to impose a specialized function on each individual. We can find many ways of accounting in part for this development, which is so important a characteristic of the modern world—the automatic complication of economic relations, the compression within the limits of the earth's surface of a living mass in process of continual expansion, and a great deal besides. External pressures of this sort undoubtedly play a part in what is happening. But taken as a whole and in its essentials the phenomenon can only be interpreted as a basic transformation, that is to say a change of major dimensions in the human state, of which comparative biology suggests the cause. The immense social disturbances which today so trouble the world appear to signify that Mankind in its turn has reached the stage, common to every species, when it must of biological necessity undergo the coordination of its elements. *In our time Mankind seems to be approaching its critical point of social organization.*

But Man is not an insect. Nothing is more pathetic than the total and blind devotion of an ant to its ant hill; and to us nothing could be more deplorable. The ant toils without respite until it dies of exhaustion in a state of complete self-detachment whose absolute nature and "faceless" purpose are precisely what we find repugnant. Are we too to sink irresistibly, victims of an inevitable process of organic determinism, into a state in which our individual personality is wholly destroyed? The thing is inconceivable. Birth and death and the reproductive function, these are common to both men and animals. But Man, because he is capable of reflection and of planning his own actions, does not blindly respond to these laws like an animal: he assimilates and transforms them, investing them with a meaning and an intelligible moral value. Our species, let us accept it, is entering its phase of socialization; we cannot continue to exist without undergoing the transformation which in one way or another will forge our multiplicity into a whole. But how are we to encounter the ordeal? In what spirit and what form are we to approach this metamorphosis so that in us it may be hominized?

This, as I see it, is the *problem of values*, deeper than any technical question of terrestrial organization, which we must all face today if we are to confront in full awareness our destiny as living beings, that is to say, our responsibilities toward "evolution." A whirlpool is beginning to appear ahead of us, in the stream which carries us along. We can already feel the first eddies and there can be no doubt that the whirlpool is far stronger than we. But, being men, we have the power of judgment to aid our navigation. I shall seek, in this paper, to pass under review the various possible courses of action open, at this critical moment, to those at the helm—that is to say, to each of us.

Finally to decide which is the best course to follow, that is the *grand option*.

2. The Possible Paths

A PRIORI (BY "dichotomic" analysis of the various outlets theoretically offered to our freedom of action) as well as *a posteriori* (by classification of the various human attitudes in fact observable around us), three alternatives, together forming a logically connected sequence, seem to express and exhaust all the possibilities open to our assessment and choice as we contemplate the future of Mankind: a) pessimism or optimism; b) the optimism of withdrawal or the optimism of evolution; c) evolution in terms of the many or of the unit.

Before we comment upon them, let us look separately at these alternatives so that we may understand their value and their relation to one another.

a Pessimism or Optimism? "Is the state of Being good or evil? That is to say, is it better to Be than not to Be?" Despite its abstract, metaphysical form, this is essentially a practical question representing the fundamental dilemma upon which every man is compelled to pronounce, implicitly or explicitly, by the very fact of having been born. Without having willed it, without knowing why, we find ourselves engaged in a world which seems to be laboriously raising itself to a state of ever greater organic complexity. This universal stream on which we are borne expresses in material terms, within the field of our experience, the preference of Nature for Being over non-Being, for life over non-life—Being and Life manifesting and evaluating themselves through the growth of consciousness. But can this instinctive choice on the part of Nature withstand the critical activity of Thought? This question could remain at the back of our minds so long as the human task did not appear to extend beyond the need of assuring as agreeable or tolerable an existence as possible for each of the individual elements of Mankind. But it comes to the forefront, it thrusts itself urgently upon us, directly Life shows signs, as it does today, of requiring us, by very virtue of its movement toward a state of higher Being, to sacrifice our individuality. There can be no doubt that the burden of continuing the World weighs more and more heavily on the shoulders of Mankind. How immense it has already become, this ever-growing task of enabling the world to live and progress! We are like the ant that slaves itself to death that its fellow slaves may live! Is not each of us therefore a dupe, a Sisyphus? For centuries a whole order of men served another, privileged order without asking whether this state of inequality was really beyond remedy; until in the end they rebelled. Is there not reason for Man, become aware of the direction in which Life is taking him, to rebel at last; to go on strike against a blind course of evolution which may not, in any event, betoken any real progress? "Time, space, becoming, Me, images of the Void. Nothing is born of anything else, and nothing is necessary to the existence of any other thing," so wrote a contemporary philosopher (A. Consentino). It is inevitable, as the collective effort required of men costs more and more, that the dilemma, already present to clear-sighted minds, should eventually disclose itself to the mass. Is the Universe utterly pointless, or are

we to accept that it has a meaning, a future, a purpose? On this
fundamental question Mankind is already virtually divided into the
two camps of those who deny that there is any significance or value
in the state of Being, and therefore no Progress; and those, on the
other hand, who believe in the possibility and the rewards of a
higher state of consciousness.

For the first only one attitude is possible: a refusal to go further;
desertion which is equivalent to turning back. For these no further
problem arises, since they are lodged in incoherence and disintegration. We may leave them there. But those in the other camp are
confronted by the call of duty and the problems of a further advance. Let us follow them toward the logical end of their position.

b Optimism of Withdrawal or Optimism of Evolution. To have decided in favor of the value of Being, to have accepted that the world
has a meaning and is taking us somewhere, does not necessarily imply that we must follow its apparent course further, or *a fortiori* to the
end. Walking through a town we often have to make a sharp turn to
right or left in order to reach our destination. Centuries ago the
wise men of India were struck by the enslaving and inescapable
character of the environment in which human activities are con-
ducted. The greater our efforts to know and possess and organize
the world, they observed, the more do we strengthen the material
trammels that imprison us and increase the universal multiplicity
from which we must free ourselves if we are to attain the blessed
unity. They concluded, therefore, that there was no conceivable way
of approach to the state of higher Being except by breaking the
bonds that confine us. We must persuade ourselves of the nonexis-
tence of all surrounding phenomena, destroy the Grand Illusion by
asceticism or by mysticism, create night and silence within ourselves; then, at the opposite extreme of appearance, we shall penetrate to what can only be defined as a total negation—the ineffable

Reality. Such is the thinking of Oriental wisdom; and there is still an appreciable number of Christians who think on similar lines, although far less radically (since their God comprises all the determinisms in which Nirvana is lacking). Seeing that a state of total socialization awaits the human species, they ask, can we fail to recognize the Eastern concept of Karma in this monstrous form? What we call civilization is weaving its web around us with a terrifying rapidity. Let us cut the threads while there is yet time. Pursuing all the paths of detachment and contemplation, not from disdain but from excessive esteem for the state of Being, let us break away from the evolutionary determinism, break the spell, withdraw.

Thus at the outset there is a cleavage in the "optimist" branch of Mankind. On the one hand there are those who see our true progress only in terms of a break, as speedy as possible, with the world: as though the spirit could not exist, or at least could not henceforth fulfill itself, except in separation from matter. And there are those on the other side, the believers in some ultimate value in the tangible evolution of things. For these latter (the true optimists), the tasks and difficulties of the present day by no means signify that we have come to an impasse in our evolution. Their faith in the Universe is stronger than any temptation to withdraw. The worst of courses, in their view, would be to retreat from the whirlpool, or alter course in order to avoid it. The way out (since this certainly exists!) can only be further ahead—forward beyond the rapids. It is in intelligent alliance with the rising tides of matter that we shall draw nearer to the Spirit.

Withdrawal, or evolution proceeding ever further? This is the second choice that human thought encounters in its search for a solution to the problem of action.

At this new point of bifurcation two attitudes are defined—two "mentalities" disclose themselves and separate. We may leave the believers in withdrawal to go their way along a road which vanishes from our sight. Let us follow the others, those who are faithful to Earth, in their effort to steer the human vessel onward through the tempests of the future. This second group may at first sight appear to be homogeneous; but in fact it is not yet wholly one in mind or spirit. A final cleavage is necessary to separate absolutely, in a pure state, the conflicting spiritual tendencies which are confusedly intermingled in the present world, at the heart of human freedom.

c *Plurality or Unity?* As we have shown, the subdivision of what one may call "the human spiritual categories" begins logically with faith in the state of Being, and proceeds to faith in the further progress of the material world around us—that is to say, in the most fundamental terms, *faith in the spiritual value of matter*. But psychologically this dichotomic process, whereby at each point of choice something like a new spiritual species breaks away, is influenced throughout by a final orientation which qualifies or obscurely dictates each of the earlier choices: "In what direction and in what form are we to look for this new state of being which we expect to be born of our future development? Is the Universe, of its nature, scattering itself in sparks; or on the contrary is it tending to concentrate in a single center of light?" Plurality or Unity? Two possibilities determining two basic attitudes, more radical than any difference of race, nationality, or even formal religion; and between them runs the true line of the spiritual division of the Earth. Pluralism or (using the word in its purely etymological sense) monism? This is the ultimate choice, by way of which Mankind must finally be divided, knowing its own mind.

In the view of the "pluralist" the world is moving in the direction of dispersal and therefore of the growing autonomy of its separate elements. For each individual the business, the duty and the interest of life consist in achieving, *in opposition to others*, his own ut-

most uniqueness and personal freedom; so that perfection, beati-
tude, supreme greatness belong not to the whole but to the least
part. By this "dispersive" view the socialization of the human mass
becomes a retrograde step and a state of monstrous servitude—*unless* we can discern in it the birth of a new "shoot" destined eventually to bring forth stronger individualities than our own. Only with
this reservation, and within these limits, is the phenomenon to be
tolerated. Collectivization in itself, no matter what form it may
take, can only be a provisional state and one of relative unimpor-
tance. Evolution culminates, by the progressive isolation of its
fibers, in each separate individual and even in each moment of the
individual's life. Essentially, as the "pluralist" sees it, the Universe
spreads like a fan: it is *divergent* in structure.

But to the "monist" the precise opposite is the case: nothing exists or finally matters except the Whole. For the elements of the
world to become absorbed within themselves by separation from
others, by isolation, is a fundamental error. The individual, if he is
to fulfill and preserve himself, must strive to break down every kind
of barrier that prevents separate beings from uniting. His is the exaltation, not of egoistical autonomy but of communion with all
others! Seen in this light the modern totalitarian regimes, whatever
their initial defects, are neither heresies nor biological regressions:
they are in line with the essential trend of "cosmic" movement. Pluralism, far from being the ultimate end of evolution, is merely a first
outspreading whose gradual shrinkage displays the true curve of
Nature's proceedings. Essentially the Universe is narrowing to a
center, like the successive layers of a cone: it is *convergent* in structure.

So the question can finally be posed: fulfillment of the world
by divergence, or fulfillment of the same world by convergence? It
seems that the final answer must lie in one or other of these two di-

rections, in the sense that anything else that has to be decided can only be of lesser importance. Our analysis of the different courses open to Man on the threshold of the socialization of his species comes to an end at this last fork in the road. We have encountered three successive pairs of alternatives offering four possibilities: to cease to act, by some form of suicide; to withdraw through a mystique of separation; to fulfill ourselves individually by egoistically segregating ourselves from the mass; or to plunge resolutely into the stream of the whole in order to become part of it.

Faced by this apparent indeterminacy of Life in ourselves, what are we to do? Shall we try to ignore the problem and continue to live on impulse and haphazard, without deciding anything? This we cannot do. The beasts of the field may trust blindly to instinct, without thereby diminishing or betraying themselves, because they have not yet *seen*. But for us, because our eyes have been opened, even though we seek hurriedly to close them, the question will continue to burn in the darkest corner of our thought. We cannot recapture the animal security of instinct. Because, in becoming men, we have acquired the power of looking to the future and assessing the value of things, we cannot do nothing, since our very refusal to decide is a decision in itself.

We cannot stand still. Four separate roads lie open to us, one back and three forward.

Which are we to choose?

3. The Choice of the Road

a *In search of a criterion.* The classification we have established is more than a flight of fancy. The four roads are not a fiction. They exist in reality and all of us know people embarked upon one or other of them. There are both pessimists and optimists around us; and among the latter there are "buddhists," "pluralists" and "monists."

Confronted by this diversity and division of human attitudes in face of a world to be abandoned or pursued, we are apt to shrug our shoulders and say, "It's all a matter of temperament." This amounts to saying that, in every sphere, faith or the lack of faith means no more and is no more controllable than a tendency of the spirit toward sadness or joy, music or geometry. A comfortable explanation, since it renders discussion unnecessary; but an inadequate one, since it purports to settle, by invoking the subjective side of our nature, a problem that is essentially objective, namely that of the structure peculiar to the world in which we find ourselves. For let us face it: to each of the four choices we have outlined there must necessarily correspond a Universe of an especial kind—disorderly or ordered, exhausted or still young, divergent or convergent. And of these four kinds of Universe *only one can exist at a time—only one is true.* We are no more free to follow our impulses blindly in the ordering of our lives than is the captain of a ship heading for a prescribed harbor. Accordingly we need some criterion of values to enable us to make our choice. But immersed in the Universe as we are, we have no means of getting outside it, even momentarily, to see if it is going anywhere, and if so where. We have no periscope; we are navigating in the depths. Is there nothing within the world to enable us to judge whether we inside it are moving in the right direction, that is to say, in the same direction as it is moving itself?

Yes, there is a clear indication, and it is the one of which we have already spoken: the growth, within and around us, of a greater consciousness. More than a century ago the physicists observed that, in the world as we know it, the fraction of *unusable* energy (entropy) is constantly increasing; and they found in this a mathematical expression of the irreversibility of the cosmos. This absolute of physics has thus far not only resisted all attempts at "relativization," but, if I am not mistaken, it tends to find its counterpart in a current moving in the opposite sense, positive and constructive, which is revealed by the study of the earth's biological past: the ascent of the Universe toward zones of increasing improbability and personality. Entropy and life; backward and forward: two complementary expressions of the arrow of time. For the purposes of human action, entropy (a mass-effect rather than a law of the unit) is without meaning. Life, on the other hand, if it is understood to be the growing interiorization of cosmic matter, offers to our freedom of choice a precise line of direction. Confronted by the phenomenon of "socialization" in which Mankind is irresistibly involved, do we seek to know how to act that we may better conform to the secret processes of the world of which we are a part? Then of the alternatives that are offered we must choose the one which seems best able to develop and preserve in us the highest degree of consciousness. If we turn out to have been wrong in this, then the Universe has no less gone astray.

b *Reduction of the alternatives.* To have accepted, on the strength of historical evidence, that the world reveals through its past its progress toward the Spirit, is to recognize equally that we need no longer choose between being and non-being. Indeed, how can we choose when we are already enrolled? The choice was made long before we were born; or more exactly, it was of the choice itself that we were born, inasmuch as the choice is implied in the progress of the Universe, that from the first has followed a preordained course. An underlying doubt as to the primacy of consciousness over unconsciousness might at a pinch be conceivable in a mind emerging suddenly from nothingness; but it seems contradictory in an evolved being whose origins attest to this primacy. In their extreme form pessimism and agnosticism are condemned by the very fact of our existence. Therefore we need not hesitate in rejecting them. This disposes of the first alternative.

The second alternative seems to pose a more delicate problem. "Withdrawal—or evolution proceeding ever further?" In which direction does a higher state of consciousness await us? Here, at first sight, the answer is less clear. There is nothing contradictory in itself in the idea of human ecstasy sundered from material things. Indeed, as we shall see, this fits in very well with the *final* demands of a world of evolutionary structure. But with one proviso: that the world in question shall have reached a stage of development so advanced that its "soul" can be detached without losing any of its completeness, as something wholly formed. But have we any reason to suppose that human consciousness *today* has achieved so high a degree of richness and perfection that it can derive nothing more from the sap of the earth? Again we may turn to history for an answer. Let us suppose, for example, that the strivings and the progress of civilization had come to an end at the time of Buddha, or in the first centuries of the Christian era. Can we believe that nothing essential, of vision and action and love, would have been lost to the Spirit of Earth? Clearly we cannot. And this simple observation alone suffices to guide our decision. So long as a fruit continues to grow and ripen we refrain from picking it. In the same way, so long as the world around us continues, even in suffering and disorder, to yield a harvest of problems, ideas and new forces, it is a sign that we must continue to press forward in the conquest of matter. Any immediate withdrawal from a world of which the burden grows heavier every day is denied to us, because it would certainly be premature. So much for the second alternative.

And so, since we are bound to press on, we find ourselves faced by the third alternative. What course are we to adopt in order that our personal efforts may most effectively contribute to the terrestrial consciousness which we must strive to heighten and extend? Is it to be a jealously guarded fostering of our own individuality, achieved in increasing isolation; or in the association and giving of ourselves to the collective whole of Mankind? Are we to reject or accept human socialization, elect for a divergent or a convergent world? Where is the truth? Which is the right way?

It appears to me that at this last fork in the road the modern problem of Action displays itself in its most essential and acute form. If there is any characteristic clearly observable in the progress of Nature toward higher consciousness, it is that this is achieved by increasing differentiation, which in itself causes ever stronger individualities to emerge. But it would seem that individualization leads to opposition and separation. In logic, therefore, we are led to suppose that every man must fight to break away from any influence that threatens to dominate and restrict him. And does not this separatist tendency exactly correspond to one of the deepest instincts of our being? But what is the voice that speaks to us in the exaltation of separateness and self-enclosure? Is it a challenge or a seduction?

It is undeniable that, viewed in a certain light, a Universe of divergent or pluralistic structure seems to be capable of giving rise to localized paroxysms of consciousness. The man who thinks to gamble the whole world for the sake of his own existence, and to gamble his own existence for the sake of the moment, is bound to live every minute with extraordinary intensity. But if we look at it we can see that this brilliance, besides being pitifully limited in scope, is radically destructive of the spirit in which it springs to light. For one thing, though it may enable the individual to achieve the heights of momentary ecstasy, it robs him in return of the ineffable joys of union and conscious loss of self in that which is greater than self: the element burns up all its future in a flying spark. And again, since the impulse must logically spread from one to another through all the elements, it becomes a process of *general volatilization* infecting Mankind as a whole. To adopt the hypothesis of a *final* divergence of Life is, in fact, to introduce biologically into the thinking part of the world an immediate principle of disintegration and death. It is to reestablish, at the very antipodes of Consciousness (become no more than a fleeting reality!), the primacy and preponderant stability of Matter. It is to deny, even more gravely than by an ill-timed act of withdrawal, the historic impulses of Life.

So there is no way out, if we wish to safeguard the preeminence of the spirit, except by taking the one road that remains to us, which leads to the preservation and further advance of consciousness—the road of unification. A convergent world, whatever sacrifice of freedom it may seem to demand of us, is the only one which can preserve the dignity and the aspirations of the living being. Therefore *it must be true.* If we are to avoid total anarchy, the source and the sign of universal death, we can do no other than plunge resolutely forward, even though something in us perish, into the melting pot of socialization.

Though something in us perish?

But where is it written that he who loses his soul shall save it?

4. *The Properties of Union*

IT IS AT this point that we must rid ourselves of a prejudice which is deeply embedded in our thought, namely the habit of mind which causes us to contrast unity with plurality, the element with the whole and the individual with the collective, as though these were diametrically opposed ideas. We constantly argue as though in each case the terms varied inversely, a gain on the one side being *ipso facto* the other side's loss; and this in turn leads to the widespread idea that *any* destiny on "monist" lines would exact the sacrifice and bring about the destruction of all personal values in the Universe.

The origin of this prejudice, which is largely imaginary, can no doubt be traced to the disagreeable sense of loss and constraint which the individual experiences when he finds himself involved in a group or lost in a crowd. It is certainly the case that any agglomeration tends to stifle and neutralize the elements which compose it; but why should we look for a model of collectivity in what is no more than an aggregate, a "heap"? Alongside these massive inorganic groupings in which the elements intermingle and drown, or more exactly at the opposite pole to them, Nature shows herself to be full of associations brought about and organically ordered by a precisely opposite law. In the case of associations of this kind (the only true and natural associations) the coming together of the separate elements does nothing to eliminate their differences. On the contrary, *it exalts them*. In every practical sphere *true union* (that is to say, synthesis) does not confound; *it differentiates*. This is what it is essential for us to understand at the moment of encountering the Grand Option.

Evidence of the fact that union differentiates is to be seen all around us—in the bodies of all higher forms of life, in which the cells become almost infinitely complicated according to the variety of tasks they have to perform; in animal associations, where the individual "polymerises" itself, one might say, according to the function it is called upon to fulfill; in human societies, where the growth of specialization becomes ever more intense; and in the field of personal relationships, where friends and lovers can only discover all that is in their minds and hearts by communicating them to one another. We may note, certainly, that in these various forms of collective life (except the last) differentiation, the fruit of union, goes hand-in-hand with mechanization, the element becoming a cog in the machine; and that this is especially what happens in the case of the termitary and the hive, of which the shadow looms so disturbingly over the collective future of Mankind. But we must take care not to bring phenomena of a different order into our argument without making the necessary adjustments. In the termitary and the hive (as in the case of the cells of our own body) the union and therefore the specialization of the elements takes place in the field of *material functions*—nutrition, reproduction, defense, etc.—which accounts for the transformation of the individual into a standardized part. But let us imagine another kind of association within which a different possibility of mutual fulfillment is offered to the individuals composing it, this time a psychic grouping corresponding to what might be called a *function of personalization*. Operating in such a field, the tendency of union to bring about differentiation, far from giving birth to a mere mechanism, must have the effect of increasing the variety of choice and the wealth of spontaneity. Anarchic autonomy tends to disappear, but it does so in order to achieve its consummation in the harmonized flowering of individual values.

And this is precisely what happens in the case of Mankind. By virtue of the emergence of Thought a special and novel environment has been evolved among human individuals within which they acquire the faculty of associating together, and reacting upon one another, no longer primarily for the preservation and continuance of the species but for the creation of a common consciousness. In such an environment the differentiation born of union may act upon that which is most unique and incommunicable in the individual, namely his personality. Thus socialization, whose hour seems to have sounded for Mankind, does not by any means signify the ending of the Era of the Individual upon earth, but far more its beginning. All that matters at this crucial moment is that the massing together of individualities should not take the form of a functional and enforced mechanization of human energies (the totalitarian principle), but of a "conspiracy" informed with love. Love has always been carefully eliminated from realist and posi-

tivist concepts of the world; but sooner or later we shall have to acknowledge that it is the fundamental impulse of Life, or, if you prefer, the one natural medium in which the rising course of evolution
can proceed. With love omitted there is truly nothing ahead of us
except the forbidding prospect of standardization and enslavement—the doom of ants and termites. It is through love and
within love that we must look for the deepening of our deepest self,
in the life-giving coming together of humankind. Love is the free
and imaginative outpouring of the spirit over all unexplored paths.
It links those who love in bonds that unite but do not confound,
causing them to discover in their mutual contact an exaltation capable, incomparably more than any arrogance of solitude, of
arousing in the heart of their being all that they possess of uniqueness and creative power.

We may have supposed when a moment ago we were bidding
farewell to a Universe of divergence and plurality, that some part
of our individual riches must be absorbed by our immersion in
Life as a whole. Now we see that it is precisely through this apparent sacrifice that we may hope to attain the high peak of personality which we thought we must renounce.

Nor is this all.

Union differentiates, as I have said; the first result being that it
endows a convergent Universe with the power to extend the individual fibers that compose it without their being lost in the whole.
But this mechanism, in such a Universe, begets another property. If
by the fundamental mechanism of union the elements of consciousness, drawing together, enhance what is most incommunicable in themselves, it means that the principle of unification causing
them to converge is in some sort a separate reality, distinct from
themselves; not a "center of resultance" born of their converging,
but a "center of dominance" effecting the synthesis of innumerable
centers culminating in itself. Without this the latter would never

come together at all. In other words, in a converging Universe each element achieves completeness, not directly in a separate consummation, but by incorporation in a higher pole of consciousness in which alone it can enter into contact with all others. By a sort of inward turn toward the Other its growth culminates in an act of giving and in excentration. What does this mean except that at this final stage there reappears the mystical "annihilation" advocated by those whom we called earlier (in discussing the second alternative) the partisans of Withdrawal. Everything now becomes clear. What the apostles of ecstasy foresaw was true. But they wished to escape in an arbitrary and, as we have said, premature fashion. They were right in their desire to be absorbed in the Other; but they did not see that this mystical night or death could only be the end and apotheosis of a process of growth. Can water boil under ordinary conditions before it has reached a temperature of 100 degrees? Before passing into the Beyond, the World and its elements must attain what may be called their "point of annihilation." And it is precisely to this critical point that we must ultimately be brought by the effort consciously to further, within and around ourselves, the movement of universal convergence!

From which, to sum up, the following situation arises.

To elect in the depths of our being for the possibility and hope of an indefinitely increasing unification of the Universe, is not merely the only course we can pursue which conforms to the evolutionary past of the world; it is the course that embraces, in its essence, *every other constructive act in which we might look for an alternative.* Not only does this road offer a positive outlet for the diminished or specialized form of consciousness—a victory dearly paid for by Life—but consciousness as a whole must follow it, with all the accumulation of riches which, at each turning point, we had thought to abandon. Which amounts to saying that the world is well made! In other words, the choice which Life requires of our considered action is a great deal less complex than at first seemed to be the case; for it is reduced to a simple choice between the first and last stages of the successive alternatives which we have been able to define: the rejection of Being, which returns us to dust, or the acceptance of Being, which leads us, by way of socialization, to faith in a Supreme Unity—opposite directions along a single road.

But if, as history suggests, there is really a quality of the inevitable in the forward march of the Universe—if, in truth, the world cannot turn back—then it must mean that individual acts are bound to follow, *in the majority and freely*, the sole direction capable of satisfying all their aspirations toward every imaginable form of higher consciousness. Having been initially the fundamental choice of the individual, the Grand Option, that which decides in favor of a convergent Universe, is destined sooner or later to become the *common choice* of the mass of Mankind. Thus a particular and generalized state of consciousness is presaged for our species in the future: a "conspiration" in terms of perspective and intention.

Which brings us in conclusion to the consideration of an especial phenomenon arising directly out of this approaching unanimity—the more or less early establishment on earth of a new atmosphere, or better, a new environment of action.

5. *The True Environment of Human Action*

THE HISTORIANS OF philosophy, in their study of the development of thought through the ages, prefer to dwell upon the birth and evolution of ideas, theses, formally constructed systems. But arguable schemes of this sort do not constitute the whole, or perhaps even the most important part, of the life of the spirit. A geometrical system is made up of points, lines and diagrams, but in the deeper sense it depends on the type of space (number of dimensions, curvature) in which the geometer operates. According to the nature of this space properties change or are generalized, and certain transformations and movements become possible. Space in itself is something that overflows any formula; yet it is in terms of this inexpressible that a whole expressible world is interpreted and developed. But what is true and clearly apparent in the abstract field of geometry may also be found, and should be examined with no less care, in the general systematization of phenomena which we call philosophy. To philosophize is to put in order the lines of reality around us. What first emerges from any philosophy is a coherent whole of harmonized relationships. But this whole, if we look closely, is always conceived in terms of a Universe intuitively endowed with certain fixed properties which are not a thing in themselves but *a general condition of knowledge*. If these properties should change, the whole philosophy, without necessarily breaking down, must adapt itself and readjust the relation between its parts; like a design on a sheet of paper which undergoes modification when the paper is curved. Indeed the past history of human intelligence is full of "mutations" of this kind, more or less abrupt, indicating, in addition to the shift of human ideas, an evolution of the "space" in which the ideas took shape—which is clearly very much more suggestive and profound.

Let me cite a single instance, the most recent, of this sort of transformation.

Until the sixteenth century men in general thought of space and time as though they were limited compartments in which objects were juxtaposed and interchangeable. They believed that a geometrical envelope could be traced round the totality of the stars. They talked, thinking they understood, of a first and last moment discernible in the past and the future. They argued as though every element could be arbitrarily moved, without changing the world, to any point along the axis of time. The human mind believed itself to be perfectly at home in this universe, within which it tranquilly wove its patterns of metaphysics. And then one day, influenced by a variety of internal and external causes, this attitude began to change. Spatially our awareness of the world was extended to embrace the Infinitesimal and the Immense. Later, in temporal terms, there came the unveiling, behind us and ahead, of the abysses of Past and Future. Finally, to complete the structure we became aware of the fact that, within this indefinite extent of space-time, the position of each element was so intimately bound up with the genesis of the whole that it was impossible to alter it at random without rendering it "incoherent," or without having to readjust the distribution and history of the whole around it. To accommodate this expansion of our thought the restricted field of static juxtaposition was replaced by a field of evolutionary organization which was limitless in all directions (except forward, in the direction of its pole of convergence). It became necessary to transpose our physics, biology and ethics, even our religion, into this new sphere, and this we are in process of doing. We can no more return to that sphere which we recently left than a three-dimensional object can enter a two-dimensional plane. The *general* and also the *irreversible* modification of perceptions, ideas, problems: these are two indications that the spirit has acquired an added dimension.

Let us now turn to the psychological effects of this Grand Option in virtue of which, as we have said, Mankind must elect to adopt a general perspective and habit of mind appropriate to its participation in a Universe of convergent consciousness. What may we expect to be the inner consequences of the change? Hitherto Man as a whole has lived practically speaking without attempting any far-going analysis of the conditions proper to and ensuing from his activities. He has lived from hand to mouth in the pursuit of more or less immediate and limited aims, more by instinct than by reason. But now the atmosphere around him becomes sustaining, consistent and warm. As he awakens to a sense of "universal unification" a wave of new life penetrates to the fiber and marrow of the least of his undertakings, the least of his desires. Everything glows, expands, is impregnated with an essential savor of the Absolute. Even more, everything is animated with a flow of Presence and of Love—the spirit which, emanating from the supreme pole of personalization, fosters and nourishes the mutual affinity of individualities in process of convergence. Will it be possible for us, having savored this climate, to turn back and tolerate any other? A *general* and *irreversible* readjustment of the values of existence: again two indications (this time not in terms of vision but in the field of action) showing our accession, beyond all ideologies and systems, to a different and higher sphere, a new spiritual dimension.

It truly seems that for Man this is the greatness of the present moment. Further ideological clashes and moral dissensions lie in wait for us as we go forward; and also further unions and further triumphs. But the succeeding acts of the drama must take place on another level; they must occur in a new world into which, at this moment, we are being born: a world in which each thinking unit upon earth will only act (if he *agrees* to act) in the consciousness, become natural and instinctive to all, of furthering a work of total personalization.

When it has passed beyond what we called at the beginning its "critical point of socialization" the mass of Mankind, let this be my conclusion, will penetrate for the first time into the environment which is biologically requisite for the wholeness of its task.

PARIS, MARCH 3, 1939. *CAHIERS DU MONDE NOUVEAU,* 1945.

CHAPTER 4

SOME REFLECTIONS
ON PROGRESS

PART I. THE FUTURE OF MAN SEEN
BY A PALEONTOLOGIST

Introduction

WHEN LITTLE MORE than a century ago, Man first discovered the abyss of time that lies behind him, and therefore the abyss that lies ahead, his first feeling was a tremendous hope, a sense of wonderment at the progress our fathers had made.

But now the wind seems to have changed. Following many setbacks a wave of troubled scepticism (adorned with the name of "realism") is sweeping through the world. Whether from immobilist reaction, sick pessimism or simply pose, it has become "good form" to deride or mistrust anything that looks like faith in the future.

"Have we ever moved? Are we still moving? And if so, are we going forward or back or simply in a circle?"

This is an attitude of doubt that will prove fatal if we do not take care, because in destroying the love of life it also destroys the life-force of Mankind.

I wish to show in this paper that, however bitter our disillusionment with human goodness in recent years, there are stronger scientific reasons than ever before for believing that we do really progress and that we can advance much further still, provided we are clear about the direction in which progress lies and are resolved to take the right road.

1. Preliminary Observations: The Slow Movements

TO UNDERSTAND WHAT follows we must first thoroughly assimilate the idea that there are movements in the Universe so slow that we cannot directly detect them. The idea of slow movement is in itself very simple and commonplace—we have all looked at the hour-hand of a watch. But it took us a long time to realize that the more stable and immobile a given object in Nature may appear to be, the greater is the likelihood that it represents a profound and majestic process of movement. We know now that the vast system of stars in our own sky is composed of a single nebula, the Milky Way, in course of granulation and deployment; and that this nebula, in association with millions of other spiral units, forms a single, immense supersystem which is also in process of expansion and organization. We know that the continents tremble and that the mountains continue to rise beneath our feet . . . and so on.

It can be said that Science today progresses only by peeling away, one after another, all the coverings of apparent stability in the world; disclosing beneath the immobility of the infinitely small, movement of extra rapidity, and beneath the immobility of the Immense, movement of extra slowness.

We are concerned here with the second of these effects, which may be expressed as follows: everything in the Universe moves; but *the larger a thing is, the slower is its movement.*

2. The Case of Life

THIS BEING POSITED let us leave the nebulae and the mountains and turn to Life itself, of which Mankind is a fragment.

Life, by our timescale, is a phenomenon of prodigious age— over 300 million years. Moreover it is composed of myriads of separate elements and it covers the earth. In terms of space-time Life comes in the category of immensely large things. It is part of the Immense, and if it moves at all it moves like the Immense.

Our object is to determine whether Life and Mankind move. We can only find out by observing them (like the hour-hand of our watch) over a *very great length of time.* Here it is that we see the part played by paleontology, as well as the secret vice of our critics.

3. The Role of Paleontology

IT MIGHT SEEM that paleontology is a science of pure speculation or inquisitiveness, and the paleontologist the most unreal and useless of researchers; a man dedicated to retrospection, plunged living into the past, where he spends his days collecting the debris of all kinds of dead things. That is certainly what many laymen think, and it may well be the view humbly taken by many paleontologists of themselves.

But in this the instinct that prompts our work sees more clearly than reason. The reconstruction of "that which was" may rationally appear to be merely a fantasy for idle minds; but in fact the meticulous work accomplished in the past hundred years by the collectors of fossils, the results which they have patiently recorded in innumerable papers and in barbarous language, perfectly incomprehensible to non-initiates, the paraphernalia of systematized knowledge and the clutter on the museum shelves, all this has made a contribution of the utmost importance to the world's thinking. It has added to the sum of human knowledge an item of extraordinary interest—*a segment of the past extending over some* 300 *million years.*

Do we fully realize its value?

We are trying, let me repeat, for vital reasons to determine whether the world, Mankind, is the seat of any kind of progress. Let us put aside all metaphysical speculation, all sentimental impressions and arguments. We are dealing with a question of fact and we must look at the facts. If we confine ourselves to short periods of time on which progress makes no mark our argument will drag on and get nowhere. But if we contemplate a depth of time such as this one that we have been able to reconstruct in the laboratory, any movement of Life, if such exists, must of necessity show itself.

Instead of arguing fruitlessly within the overbrief space of a few generations, let us look at the broad vista which science offers us. What do we see?

4. *The Growth of Consciousness*

FOR VARIOUS PSYCHOLOGICAL and technical reasons which I need not examine here, the reading or decipherment of the tract of time disclosed by paleontology is still not free of difficulty. Indeed it continues to be a matter of vehement dispute. The interpretation which I am about to put forward must therefore not be regarded as "accepted." Nevertheless it seems to me so self-evident that I have no hesitation in offering it as the correct interpretation and the one destined sooner or later to win general scientific agreement.

It may be stated thus: when observed through a sufficient depth of time (millions of years) Life can be seen to move. Not only does it move but it advances in a definite direction. And not only does it advance, but in observing its progress we can discern the process or practical mechanism whereby it does so.

These are three propositions which may be briefly developed as follows.

a *Life moves.* This calls for no demonstration. Everyone in these days knows how greatly all living forms have changed if we compare two moments in the earth's history sufficiently separated in time. In any period of ten million years Life practically grows a new skin.

b *In a definite direction.* This is the crucial point which has to be clearly understood. While accepting the undeniable fact of the general evolution of Life in the course of time, many biologists still maintain that these changes take place without following any defined course, in any direction and at random. This contention, disastrous to any idea of progress, is refuted, in my view, by the tremendous fact of the continuing "cerebralization" of living creatures. Research shows that from the lowest to the highest level of the organic world there is a persistent and clearly defined thrust of animal forms toward species with more sensitive and elaborate nervous systems. A growing "innervation" and "cephalization" of organisms: the working of this law is visible in every living group known to us, the smallest no less than the largest. We can follow it in insects as in vertebrates; and among the vertebrates we can follow it from class to class, from order to order, and from family to family. There is an amphibian phase of the brain, a reptilian phase, a mammalian phase. In mammals we see the brain grow as time passes and become more complex among the ungulates, the carnivores and above all the primates. So much so that one could draw a steadily rising Curve of Life taking Time as one coordinate and, as the other, the quantity (and quality) of nervous tissue existing on earth at each geological stage.

What else can this mean except that, as shown by the development of nervous systems, there is a continual heightening, a rising tide of consciousness which visibly manifests itself on our planet in the course of the ages?

c We come to the third point. What is the *underlying process* whose existence we can perceive in this continual heightening of consciousness, as revealed by the organic evolution of the nervous system and the brain? Let us look more closely in the light of the latest data supplied by the combined ingenuity of an army of research workers. As we are beginning to realize, there are probably tens of thousands of atoms grouped in a single virus molecule. There are certainly tens of thousands of molecules grouped in a single cell. There are millions of brains in a single ant hill. . . .

What does this atomism signify except that Cosmic Matter, governed at its lower end (as we already know) by forces of dispersal which slowly cause it to dissolve into atoms, now shows itself to be subjected, at the other end, to an extraordinary power of enforced coalescence, of which the outcome is the emergence, *pari passu*, of an ever-increasing amount of spiritual energy in matter that is ever more powerfully synthesized? Let me note that there is nothing metaphysical in this. I am not seeking to define either Spirit or Matter. I am simply saying, without leaving the physical field, that the greatest discovery made in this century is probably the realization that the passage of Time may best be measured by the gradual gathering of Matter in superposed groups, of which the arrangement, ever richer and more centralized, radiates outward from an ever more luminous fringe of liberty and interiority. The phenomenon of growing consciousness on earth, in short, is directly due to the increasingly advanced organization of more and more complicated elements, successively created by the working of chemistry and of Life. At the present time I can see no more satisfactory solution of the enigma presented to us by the physical progress of the Universe.

5. *The Place of Man in the Forefront of Life*

IN WHAT I have said thus far I have been looking at Life in general, in its entirety. We come now to the particular case which interests us most—the problem of Man.

The existence of an ascendant movement in the Universe has been revealed to us by the study of paleontology. Where is Man to be situated in this line of progress?

The answer is clear. If, as I maintain, the movement of the cosmos toward the highest degree of consciousness is not an optical illusion, but represents the essence of biological evolution, then, in the curve traced by Life, Man is unquestionably situated at the topmost point; and it is he, by his emergence and existence, who finally proves the reality and defines the direction of the trajectory—"the dot on the i". . . .

Indeed, within the field accessible to our experience, does not the birth of Thought stand out as a critical point through which all the striving of previous ages passes and is consummated—the critical point traversed by consciousness, when, by force of concentration, it ends by reflecting upon itself?

Prior to Galileo, science thought of Man as the mathematical and moral center of a World composed of spheres turning statically upon themselves. But in terms of our modern neoanthropocentricity, Man, both diminished and enlarged, becomes the *head* (terrestrial) of a Universe that is in the process of psychic transformation—Man, the last-formed, most complex and most conscious of "molecules." From which it follows that, borne on the tide of millions of years of psychogenesis, we have the right to consider ourselves the fruit of a progression—the children of progress.

The world did at least progress to the point where the firstborn of our race appeared. Here we have a fixed and solid point on which to base our philosophy of life.

Let us now take a further step.

We may agree that zoological evolution culminated in Man. But having reached this peak did it come to a stop? Life continued to move until Thought entered the world, this we may admit. But has it advanced since then? Can it make any further progress?

6. *The Movement of Mankind upon Itself*

ANCIENT THOUGH PREHISTORY may make it seem to our eyes, Mankind is still very young. We can trace its existence for not much more than a hundred thousand years, a period so short that it has left no mark on the majority of the animal forms that preceded us on the earth and which still surround us. It may seem impossible, and it is certainly a very delicate matter, to measure any movement of Life in so slender a fragment of the past. Nevertheless, owing to the exceptionally rapid development which is a characteristic of the human wave, a direct assessment of the advance of our own group in terms of consciousness is possible to the practiced eye, even within this limited tract of time.

a It seems in the first place that, anatomically, a gradual evolution of the brain can be discerned during the earliest phases of our phylogenesis. Pithecanthropus and Sinanthropus possessed intelligence, but there are solid grounds for supposing that they were not cerebrally as well developed as ourselves.

b We may accept that the human brain reached the limit of its development at the stage which anthropologists call *Homo sapiens*; or at least, if it has continued to develop since then, that the change cannot be detected by our present methods of observation. But although, since the Age of the Reindeer (that is to say, within a period of twenty or thirty thousand years) no progress is perceptible in either the physical or the mental faculties of Individual Man, the fact of organopsychic development seems to be clearly manifest in Collective Man: and this, whatever we may think of it, represents as true an advance as the acquisition of an added convolution by the brain.

Let me here repeat the two fundamental equations or equivalents which we have established:

Progress = growth of consciousness.
Growth of Consciousness = effect of organization.

Taken together these mean that, in order to discover or verify the existence of biological progress within a given system, we have only to observe, for the period of time and the field we are considering, how far the state of organization varies within that system.

This being posited we may compare the world of the cave dweller with the world of today. Setting all theory aside there can be no question but that, within this period of thirty thousand years, Mankind has advanced almost unbelievably in its state of concentration.

Economic concentration, manifest in the unification of the earth's energies.

Intellectual concentration, manifest in the unification of our knowledge in a coherent system (science).

Social concentration, manifest in the unification of the human mass as a thinking whole.

To those who have not studied its implications, this slow and irresistible flow of our history in the direction of more and more unified groupings has no particular meaning; they relegate it to the trivial category of surface and incidental phenomena, no more. But to the enlightened eye this human development, succeeding all the twists and turns of prehuman consciousness, assumes a dazzling significance. *For the two curves are a prolongation one of the other.* Tremendous events such as those through which we are now passing are seen to take shape, and with a brilliant clarity. This tremendous war which so afflicts us, this remolding, this universal longing for a new order, what are they but the shock, the tremor and the crisis, beyond which we may glimpse a more synthetic organization of the human world? And this new order, the thought of which is in all our minds, what form can it take other than a higher degree of self-awareness on the part of a Mankind become at once more complex and more centered upon itself?

No, truly: Life in emerging into Thought did not come to a stop. Not only has it moved and progressed from the protozoa to Man, but since the coming of Man it has continued to advance along its most essential path. We can feel it at this moment quivering beneath our feet! The ship that bears us is still making headway.

And it is here that the ultimate and decisive question arises, finally the only question that interests us. Thus far Life, and Man himself, has progressed. So be it. But what of the future? We are still moving, but can we continue much longer to advance?

Have we not reached a dead end? Can we talk seriously of a future for Mankind?

7. The Future of Mankind

I MAKE NO claim to be a prophet. Moreover I know, as a scientist, how dangerous it is to extend a curve beyond the facts, that is to say, to extrapolate. Nevertheless I believe that, basing the argument upon our general knowledge of the world's history over a period of 300 million years, we can advance the following two propositions without losing ourselves in a fog of speculation:

a Firstly, Mankind still shows itself to possess a *reserve*, a formidable potential of concentration, i.e., of progress. We have only to think of the immensity of the forces, ideas and human beings that have still to be born or discovered or applied or synthesized. . . . "Energetically" as well as biologically the human group is still young, still fresh. If we are to judge by what history teaches us about other living groups, it still has, organically speaking, some millions of years in which to live and develop.

b Everything leads us to believe that it really does dispose of this vast reservoir of time, which is necessary for the normal achievement of its evolution. The earth is far from having completed its sidereal evolution. We may envisage all kinds of mischance (disaster or disease) which might in theory put an end to our evolutionary progress: but the fact remains that for 300 million years Life has paradoxically flourished in the Improbable. Does not this suggest that its advance may be sustained by *some sort of complicity on the part of the "blind" forces of the Universe*—that is to say, that it is inexorable?

The more we ponder these matters the more must we realize that, scientifically speaking, the real difficulty presented by Man is not the problem of whether he is a center of constant progress: it is far more the question of how long this progress can continue, at the speed at which it is going, without Life blowing up upon itself or causing the earth on which it was born to explode. Our modern world was created in less than ten thousand years, and in the past two hundred years it has changed more than in all the preceding millennia. Have we ever thought of what our planet may be like, psychologically, in a million years' time? It is finally the Utopians, not the "realists," who make scientific sense. They at least, though their flights of fancy may cause us to smile, have a feeling for the true dimensions of the phenomenon of Man.

8. *The Advance*

HAVING CLARIFIED OUR ideas, let us see what action they require of us. If progress is to continue, it will not do so of its own accord. *Evolution, by the very mechanism of its syntheses, charges itself with an ever-growing measure of freedom.*

If indeed an almost limitless field of action lies open to us in the future, what shall our moral dispositions be, as we contemplate this march ahead?

I can think of two, which may be summarized in six words: *a great hope held in common.*

a First, the hope. This must spring to life spontaneously in every generous spirit faced by the task that awaits us; and it is also the essential *impulse*, without which nothing can be done. A passionate longing to grow, to be, is what we need. There can be no place for the poor in spirit, the sceptics, the pessimists, the sad of heart, the weary and the immobilists. Life is ceaseless discovery. Life is movement.

b A hope held in common. Here again the history of Life is decisive. Not all directions are good for our advance: one alone leads upward, that which through increasing organization leads to greater synthesis and unity. Here we part company with the whole-hearted individualists, the egoists who seek to grow by excluding or diminishing their fellows, individually, nationally or racially. Life moves toward unification. Our hope can only be realized if it finds its expression in greater cohesion and greater human solidarity.

This double point is finally established by the verdict of the Past.

9. The Crossroads

BUT HERE THERE is a grave uncertainty to be resolved. The future, I have said, depends on the courage and resourcefulness which men display in overcoming the forces of isolationism, even of repulsion, which seem to drive them apart rather than draw them together. How is the drawing together to be accomplished? How shall we so contrive matters that the human mass merges in a single whole, instead of ceaselessly scattering in dust?

A priori, there seem to be two methods, two possible roads.

a The first is a process of tightening up in response to external pressures. We are in any case inescapably subject to this through the negative action of terrestrial causes. The human mass, because on the confined surface of this planet it is in a state of continuous additive growth, in numbers and interconnections, must automatically become more and more tightly concentrated upon itself. To this formidable process of natural compression there may well be added the artificial constraint imposed by a stronger human group upon a weaker; we have only to look about us at the present time to see how this idea is seeking, indeed rushing toward, its realization.

b But there is another way. This is that, *prompted by some favoring influence,* the elements of Mankind should succeed in making effective a profound force of mutual attraction, deeper and more powerful than the surface-repulsion which causes them to diverge. Forced upon one another by the dimensions and mechanics of the earth, men will purposefully bring to life a common soul in this vast body.

Unification by external or by internal force? Compulsion or Unanimity?

I spoke earlier of the present war. Does it not precisely express the tension and interior dislocation of Mankind shaken to its roots as it stands at the crossroads, faced by the need to decide upon its future?

10. *The Choice*

GLORIOUSLY SITUATED BY life at this critical point in the evolution of Mankind, what ought we to do? We hold Earth's future in our hands. What shall we decide?

In my view the road to be followed is clearly revealed by the teaching of all the past.

We can progress only by uniting: this, as we have seen, is the law of Life. But unification through coercion leads only to a superficial pseudo-unity. It may establish a mechanism, but it does not achieve any fundamental synthesis; and in consequence it engenders no growth of consciousness. It materializes, in short, instead of spiritualizing. Only unification through unanimity is biologically valid. This alone can work the miracle of causing heightened personality to emerge from the forces of collectivity. It alone represents a genuine extension of the psychogenesis that gave us birth.

Therefore it is inwardly that we must come together, and in entire freedom.

But this brings us to the last question of all. To create this unanimity we need the bond, as I said, the cement of a favoring influence. Where shall we look for it; how shall we conceive of this principle of togetherness, this soul of the Earth?

Is it to be in the development of a common *vision*, that is to say, the establishment of a universally accepted body of knowledge, in which all intelligences will join in knowing the same facts interpreted in the same way?

Or will it rather be in common *action*, in the determination of an Objective universally recognized as being so desirable that all activity will naturally converge toward it under the impulse of a common fear and a common ambition?

These two kinds of unanimity are undoubtedly real, and will, I believe, have their place in our future progress. But they need to be complemented by something else if they are not to remain precarious, insufficient and incomplete. A common body of knowledge brings together nothing but the geometrical point of intelligences. A common aspiration, no matter how ardent, can only touch individuals indirectly and in an impersonal way that is depersonalizing in itself.

It is not a *tête-à-tête* or a *corps-à-corps* that we need; it is a heart-to-heart.

This being so, the more I consider the fundamental question of the future of the earth, the more it appears to me that the generative principle of its unification is finally to be sought, not in the sole contemplation of a single Truth or in the sole desire for a single Thing, but in the common attraction exercised by a single *Being*. For on the one hand, if the synthesis of the Spirit is to be brought about in its entirety (and this is the only possible definition of progress) it can only be done, in the last resort, through the meeting, *center to center,* of human units, such as can only be realized in a universal, mutual love. And on the other hand there is but one possible way in which human elements, innumerably diverse by nature, can love one another: it is by knowing themselves all to be centered upon a single "supercenter" common to all, to which they can only attain, each at the extreme of himself, through their unity.

"Love one another, recognizing in the heart of each of you the same God who is being born." Those words, first spoken two thousand years ago, now begin to reveal themselves as the essential structural law of what we call progress and evolution. They enter the scientific field of cosmic energy and its necessary laws.

Indeed, the more I strive, in love and wonder, to measure the huge movements of past Life in the light of palaeontology, the more I am convinced that this majestic process, which nothing can arrest, can achieve its consummation only in becoming Christianized.[1]

---

[1] Unpublished. Peking, February 22, 1941. Lecture delivered at the French Embassy, on the third of March of the same year.

PART II. ON THE POSSIBLE BASES OF
A UNIVERSAL HUMAN CREED

THE PURPOSE OF the New York meetings, if I understand it aright, is not merely to seek a superficial reconciliation between the diverse forms of Faith which divide the human spirit and make it at odds with itself, but to find what they have fundamentally in common. We seek a new spirit for a new order.

I beg to be allowed to offer a brief contribution and personal testimony, the fruit of thirty years spent in close and sincere contact with scientific and religious circles in Europe, America and the Far East.

1. The Precise Point of Divergence . . . God or the World?

IT SEEMS TO me clear above all else, setting aside the countless minor divergences, and ignoring the dull, inert mass of those who believe in nothing at all, that the spiritual conflict afflicting Mankind today arises out of the division of minds and hearts into the two profoundly separated categories of:

a Those whose hopes are directed toward a spiritual state or an absolute finality situated beyond and outside this world; *b* Those who hope for the perfection of the tangible Universe within itself.

The first of these groups, by far the older, is preeminently represented in these days by the Christians, protagonists of a transcendent and personal God.

The second group, comprising those who for a variety of reasons have dedicated their lives to the service of a Universe which they conceive as eventually culminating in some form of impersonal and immanent Reality, is of very recent origin. Throughout human history this conflict between the "servants of Heaven" and the "servants of earth" has gone on; but only since the birth of the idea of Evolution (in some sort divinizing the Universe) have the devotees of earth bestirred themselves and made of their worship a true form of religion, charged with limitless hope, striving and renunciation.

Are we to disdain the world and put it behind us, or live in it in order to master and perfect it? Mankind is rent asunder at this moment by these two concepts or rival mysticisms; and in consequence its vital power of adoration is disastrously weakened.

Such in my view is the nature of the crisis, more profound than any economic, political or social struggle, through which we are passing.

2. A Principle of Convergence: The Concept of Noogenesis

ANY TWO FORCES, provided both are positive, must *a priori* be capable of growth by merging together. Faith in God and faith in the World: these two springs of energy, each the source of a magnificent spiritual impulse, must certainly be capable of effectively uniting in such a way as to produce a resulting upward movement. But in practical terms where are we to look for the principle and the generative medium which will bring about this most desirable evolutionary step?

I believe that the principle and the medium are to be found in the idea, duly "realized," that there is in progress, within us and around us, a continual heightening of consciousness in the Universe.

For a century and a half the science of physics, preoccupied with analytical researches, was dominated by the idea of the dissipation of energy and the disintegration of matter. Being now called upon by biology to consider the effects of synthesis, it is beginning to perceive that, parallel with the phenomenon of corpuscular disintegration, the Universe historically displays a second process as generalized and fundamental as the first: I mean that of the gradual concentration of its physicochemical elements in nuclei of increasing complexity, each succeeding stage of material concentration and differentiation being accompanied by a more advanced form of spontaneity and spiritual energy.

The outflowing flood of Entropy equalled and offset by the rising tide of a Noogenesis! . . .

The greater and more revolutionary an idea, the more does it encounter resistance at its inception. Despite the number and importance of the facts that it explains, the theory of Noogenesis is still far from having established itself as a stronghold in the scientific field. However, let us assume that, as all the observable evidence suggests, it will succeed before long in gaining in one form or another the place it deserves at the head of the structural laws of our Universe. Plainly the first result will be precisely to bring about the *rapprochement* and automatic convergence of the two opposed forms of worship into which, as I said, the religious impulse of Mankind is at present divided.

Once he has been brought to accept the reality of a Noogenesis, the believer in this World will find himself compelled to allow increasing room, in his vision of the future, for the values of personalization and transcendency. Of Personalization, because a Universe in process of psychic concentration is *identical* with a Universe that is acquiring a personality. And a transcendency because the ultimate stage of "cosmic" personalization, if it is to be supremely consistent and unifying, cannot be conceived otherwise than as having emerged by its summit from the elements it super-personalizes as it unites them to itself.

On the other hand, the believer in Heaven, accepting this same reality of a cosmic genesis of the Spirit, must perceive that the mystical evolution of which he dreams presupposes and consecrates all the tangible realities and all the arduous conditions of human progress. If it is to be superspiritualized in God, must not Mankind first be born and grow *in conformity with the entire system* of what we call "evolution"? Whence, for the Christian in particular, there follows a radical incorporation of terrestrial values in the most fundamental concepts of his Faith, those of Divine Omnipotence, detachment and charity. First, Divine Omnipotence: God creates and shapes us through the process of evolution: how can we suppose, or fear, that He will arbitrarily interfere with the very means whereby He fulfills His purpose? Then, detachment: God awaits us when the evolutionary process is complete: to rise above the World, therefore, does not mean to despise or reject it, but to pass through it and sublime it. Finally, charity: the love of God expresses and crowns the basic affinity which, from the beginnings of Time and Space, has drawn together and concentrated the spiritualizable elements of the Universe. To love God and our neighbor is therefore not merely an act of worship and compassion superimposed on our other individual preoccupations. For the Christian, if he be truly Christian, it is Life itself, Life in the integrity of its aspirations, its struggles and its conquests, that he must embrace in a spirit of togetherness and personalizing unification with all things.

The sense of the earth opening and exploding upward into God; and the sense of God taking root and finding nourishment downward into Earth. A personal, transcendent God and an evolving Universe no longer forming two hostile centers of attraction, but entering into hierarchic conjunction to raise the human mass on a single tide. Such is the sublime transformation which we may with justice foresee, and which *in fact* is beginning to have its effect upon a growing number of minds, freethinkers as well as believers: the idea of a spiritual evolution of the Universe. The very transformation we have been seeking!

3. A New Soul for a New World: Faith Renewed in the Progress of Mankind

FROM THIS STANDPOINT it is at once apparent that, to unify the living forces of humanity, at present so painfully at odds, the direct and effective method is simply to sound the call-to-arms and form a solid block of all those, whether of the right or the left, who believe that the principal business of present-day Mankind is to achieve a breakthrough straight ahead by forcing its way over the threshold of some higher level of consciousness. Whether Christian or non-Christian, the people inspired by this particular conviction constitute a homogeneous category. Though they may be situated at the two extreme wings of Mankind on the march, they can advance unequivocally side by side because their attitudes, far from being mutually exclusive, are virtually an extension one of the other and ask only to be completed. What more do they need that they may know and love one another? The *union sacrée*, the Common Front of all those who believe that the World is still advancing: what is this but the active minority, the solid core around which the unanimity of tomorrow must harden?

Despite the wave of skepticism which seems to have swept away the hopes (too ingenuous, no doubt, and too materialistic) on which the nineteenth century lived, faith in the future is not dead in our hearts. Indeed, it is this faith, deepened and purified, which must save us. Not only does the idea of a possible raising of our consciousness to a state of superconsciousness show itself daily, in the light of scientific experience, to be better founded and psychologically more necessary for preserving in Man his will to act; but furthermore this idea, carried to its logical extreme, appears to be the only one capable of paving the way for the great event we look for—the manifestation of a unified impulse of worship in which will be joined and mutually exalted both a passionate desire to conquer the World and a passionate longing to be united with God: the vital act, specifically new, corresponding to a new age in the history of Earth.

I am convinced that finally it is upon the idea of progress, and faith in progress, that Mankind, today so divided, must rely and can reshape itself.

REMARKS ON A NEW YORK CONGRESS OF SCIENCE AND RELIGION.
UNPUBLISHED. PEKING, MARCH 30, 1941.

CHAPTER 5

THE NEW SPIRIT

Introduction

DURING RECENT YEARS I have sought in a long
series of essays,[1] not to philosophize in the Absolute,
but as a naturalist or physicist to discover a general
significance in the events in which we are materially
involved. A great many internal and external portents (political and social upheaval, moral and religious unease) have caused us all to feel, more or less
confusedly, that something tremendous is at present
taking place in the world. But what is it?

What I wish to offer here is the outcome of my
own thinking, expressed in a simple and clarified
form so that everyone may be able to understand
it without ambiguity, and may criticize and (this is
my great hope) correct and amplify it.

The present state of the world seems to me to
be substantially determined and explained by the
influence of two progressive changes affecting human consciousness at the deepest level.

---

[1] *Le Milieu Divin, L'Esprit de la Terre, Comment je crois, L'Énergie
humaine, L'Univers personnel, The Phenomenon of Man,* etc.

The first change, already far advanced, is taking place in the field of our vision of the world. It amounts to the acquirement by the human mind of a new faculty, the perception of Time; or more precisely the perception of what I would call "the conic curvature of Time."

The second change, related to the first but less advanced, directly affects our Action. It arises out of the gradual adjustment of human values in terms of this reappraisal of Time.

1. The Cone of Time.

2. The "conic" transposition of Action.

I deal with these in separate sections.

I. THE CONE OF TIME

1. *The Organic Depth of Time and of the Spirit*

TO UNDERSTAND THE spiritual events which are so convulsing the age we live in we need to be constantly looking back (I shall repeat this) to their common origin—the discovery of Time.

At first sight the concept of Time appears so complete in its simplicity that one wonders how it can possibly be modified or improved upon. Is it not one of the solid facts on which our consciousness is based? Yet we have only to glance over the past two centuries to see that within these few generations our temporal view of the world has come to differ greatly from that of our ancestors.

This does not mean that men had to wait till the nineteenth century before seeing how events, grouped in long series, were absorbed into the past. They talked of Time long before our day, and even measured it, so far as their instruments permitted, as we do now. But Time remained for them a homogeneous quantity, capable of being divided into parts. The course of centuries lying ahead and behind us could be conceived of in theory as abruptly stopping or beginning at a given moment, the real and total duration of the Universe being supposed not to exceed a few thousand years. On the other hand, it appeared that within those few millennia any object could be arbitrarily displaced and removed to another point without undergoing any change in its environment or in itself. Socrates could have been born in the place of Descartes, and vice versa. Temporally (no less than spatially) human beings were regarded as interchangeable.

This, broadly, is what was accepted by the greatest minds up to and including Pascal.

But since then, under the influence, unconcerted but convergent, of the natural, historical and physical sciences, an entirely new concept has almost imperceptibly shaped itself in our minds.

We have in the first place realized that every constituent element of the world (whether a being or a phenomenon) has of necessity emerged from that which preceded it—so much so that it is as physically impossible for us to conceive of a thing in Time without "something before it" as it would be to imagine the same thing in Space without "something beside it." In this sense every particle of reality, instead of constituting an approximate point in itself, extends from the previous fragment to the next in an indivisible thread running back into infinity.

Secondly we have found that the threads or chains of elements thus formed are not homogeneous over their extent, but that each represents a naturally ordered series in which the links can no more be exchanged than can the successive states of infancy, adolescence, maturity and senility in our own lives.

Finally, we have gradually come to understand that no elemental thread in the Universe is wholly independent in its growth of its neighboring threads. Each forms part of a sheaf; and the sheaf in turn represents a higher order of thread in a still larger sheaf—and so on indefinitely. So that, Time acting on Space and incorporating it within itself, the two together constitute a single progression in which Space represents a momentary section of the flow which is endowed with depth and coherence by Time.

This is the organic whole of which today we find ourselves to be a part, without being able to escape from it. On the one hand, following an interlinked system of lines of indefinite length, the Stuff of the Universe spreads and radiates outwardly from ourselves, without limit, spatially from the Immense to the Infinitesimal and temporally from the abyss of the past to the abyss of the future. On the other hand, in this endless and indivisible network, everything has a particular position defined by the development (free or predetermined) of the entire system in movement. Whereas for the last two centuries our study of science, history and philosophy has appeared to be a matter of speculation, imagination and hypothesis, we can now see that in fact, in countless subtle ways, the concept of Evolution has been weaving its web around us. We believed that we did not change; but now, like infants whose eyes are opening to the light, we are becoming aware of a world in which neo-Time, organizing and conferring a dynamic upon Space, is endowing the totality of our knowledge and beliefs with a new structure and a new direction.

Before studying the implications of this, we must look more closely at the nature and properties of the new environment into which we are being born.

2. *The Convergence of Organic Time and the Upward Growth of the Spirit*

WITHIN THE LIMITS I have outlined, our new awareness of Time may now be regarded as an accomplished fact. Excepting a few ultraconservative groups, it would not occur to any present-day thinker or scientist—it would be psychologically inadmissible and impossible—to pursue a line of thought which ignores the concept of a world in evolution.

But if the Space-Time continuum is now generally accepted as the only framework within which our thought can continue to progress, it becomes the more necessary that we should agree upon the nature and general direction of the flow on which we are borne. Is it a closed vortex, an indefinite spiral, a spreading explosion? . . . What is it that has us in its grip? Moreover, immersed in its movement as we are, do we possess any point of perspective from which we may see in what direction the cosmic stream is bearing us?

The majority of people personally known to me still regard the direction and purpose of Evolution as a riddle that is scientifically unanswerable.

But it is here, in my view, that the importance becomes manifest of an intuitive notion which, timidly evolved less than fifty years ago by a small group of human minds, is now beginning to pervade twentieth-century thought as rapidly as did the idea of evolution in the nineteenth century. The discovery of the great phenomena buried in the past opened our fathers' eyes to the vague, generalized perception of a process of evolution of Life on earth. To gain a clearer idea of the precise nature of this vast biological movement, is it not enough for us simply to open our eyes (are we not already beginning to do so?) to the extraordinary and present greatness of the phenomenon of Man?

I believe this to be the case, and I wish to show why.

It seemed, following the revolutionary ideas of first Galileo and then Lamarck and Darwin, that for the "lord of creation" little was left of his past grandeur. The demolishing of the geocentric theory, leading two centuries later to the end of anthropocentrism, left Man to think of himself as finally submerged and flattened by the "temporal" flow which his intelligence had discovered. But now he seems to be again emerging in the forefront of Nature. Evolution, so they said at the end of the last century, has simply swallowed Man up, since we have proved that it extends to Man. But observing the progress of science during recent years we can see that what is happening suggests precisely the opposite. Far from being swallowed up by Evolution, Man is now engaged in transforming our earlier idea of Evolution in terms of himself, and thereafter plotting its new outline.

Let me explain.

The three characteristics which make the human individual a truly unique object in the eyes of Science, once we have made up our minds to regard Man not merely as a chance arrival but as an integral element of the physical world, are as follows:

a an extreme physicochemical complexity (particularly apparent in the brain) which permits us to consider him the most highly synthesized form of matter known to us in the Universe;

b arising out of this, an extreme degree of organization which makes him the most perfectly and deeply centered of all cosmic particles within the field of our experience;

c finally, and correlative with the above, the high degree of psychic development (reflection, thought) which places him head and shoulders above all other conscious beings known to us.

To these may be added a fourth particularity which is also of great significance: that of being the latest product of Evolution.

It is difficult to consider these four attributes, relating them to Space-Time, without becoming aware of a prospect which, however we may seek to describe it, comes essentially to this:

Science has lately been very much preoccupied with the changing properties of Matter as we follow it in either of the two spatial directions—toward the Infinitesimal or toward the Immense. Yet progress in either direction does not bring us a step nearer to the explanation of Life. Why should we not make room in our physics for the organic axis of Time? Following this axis in the downward direction of entropy we find that matter becomes diffused and energy is neutralized. This is something that we have long known. But why should we not take into specific account the cosmic movement operating in the reverse sense, toward the higher forms of synthesis, which is so strikingly apparent? Beneath our eyes, extending from the electron to Man by way of the proteins, viruses, bacteria, protozoa and metazoa, a long chain of composites is forming and unfolding, eventually attaining an astronomical degree of complexity and arrangement, and centered *pari passu* upon itself while at the same time it animates itself. Why should we not simply define Life as the specific property of Matter, the Stuff of the Universe, carried by evolution into the zone of highest complexity? And why not define Time itself as precisely the rise of the Universe into those high latitudes where complexity, concentration, centration and consciousness grow and increase, simultaneously and correlatively?

A cosmogenesis embracing and expanding the laws of our individual ontogenesis on a universal scale, in the form of Noogenesis: a world that is *being born* instead of a world that *is*: that is what the phenomenon of Man suggests, indeed compels us to accept, if we are to find a place for Man in this process of evolution in which we are obliged to make room for him.

We still hesitate, as I have said, over the form which we may conveniently attribute to Space-Time. But the fact is that we have no more time for quibbling. If it is to be adjusted to Man, the high point and effective spearhead of evolution; if it is to contain and propagate the Noogenesis through which the march of events expresses itself with an increasing clarity, Space-Time must be given whatever form is most appropriate. Caught within its curve the layers of Matter (considered as separate elements no less than as a whole) tighten and converge in Thought, by synthesis. Therefore it is as a cone, in the form of a cone, that it can best be depicted.

And it is within this cone, newly shaped in our consciousness, in terms of it and in accordance with its requirements, that we must look to see how the transposition of all human values is irresistibly proceeding.

II. THE "CONIC" TRANSPOSITION OF ACTION

1. Toward a New Humanism

TO ACCEPT THAT Space-Time is convergent in its nature is equally to admit that Thought on earth has not achieved the ultimate point of its evolution.

Indeed, if in virtue of its especial curvature the Universe, following the line of its principal axis, is really moving toward a state of maximal synthesis; and if furthermore, as practical observation shows, its human particles, taken as a whole, still possess a formidable potential of synthesis: then our present situation cannot be anything but "energetically" unstable. We cannot stay where we are at present, either physically or psychically; but looking far ahead we may descry an ultimate state in which, organically associated with one another (*more closely* than the cells of a single brain) we shall form in our entirety a single system, ultracomplex and, in consequence, ultracentered. . . . We thought that we had reached the limit of ourselves. Now we see Mankind extending within the cone of Time beyond the individual; it coils in collectively upon itself above our heads, in the direction of some sort of higher Mankind.

Let us enumerate and assess the changes of outlook and attitude that are inescapable for any person who has become aware of this prospect. I maintain that for such as he the Universe emerges from the shadows. It shows its true face, acquires its true value, glows with a new warmth and finally is illumined from within.

Let us look rapidly, one by one, at these phases of the transformation.

a   Firstly, the Universe emerges from the shadows. That is to say, it clarifies itself to the eye of reason, and precisely in those regions where it threatened to plunge most deeply into darkness. On the one hand the overwhelming vastness of the Cosmos need no longer appall us, since the indefinite layers of Time and Space, far from being the lifeless desert in which we seemed to be lost, show themselves to be the bosom which gathers together the separate fragments of a huge Consciousness in process of growth. On the other hand Evil, in all its forms—injustice, inequality, suffering, death itself—ceases theoretically to be outrageous from the moment when, *Evolution becoming a Genesis*, the immense travail of the world displays itself as the inevitable reverse side—or better, the condition—or better still, the price—of an immense triumph. And in its turn Earth, that microscopic planet on which we are crushed together, is seen to be no longer the meaningless prison in which we thought we must suffocate: for if its limits were less narrow and impenetrable could it be the matrix in which our unity is being forged?

b Secondly, the Universe shows its true face: that is to say, it traces its outline for our liberated gaze. In its present state Morality offers a painful spectacle of confusion. Apart from a few elementary laws of individual justice, empirically established and blindly followed, who can say what is good and what is evil? Can we even maintain that Good and Evil exist while the evolutionary course on which we are embarked has no clear direction? Is striving really a better thing than enjoyment, disinterest better than self-interest, kindness better than compulsion? Lacking a lookout point in the Universe, the most sharply opposed doctrines on these vital matters can be plausibly defended. Meanwhile human energy, being without orientation, is lamentably dissipated upon earth. But this disorder comes logically to an end, all the agitation is polarized, directly the spiritual reality of Mankind is revealed, above and ahead of each human being, at the apex of the Cone of Time. The best way of reaching this objective has still to be found. But is it not in itself a consolation and a source of strength to know that Life has an objective; and that the objective is a summit; and that this summit, toward which all our striving must be directed, can only be attained by our drawing together, all of us, more and more closely and in every sense—individually, socially, nationally and racially?

c Thirdly, the Universe acquires its true value: that is to say, it grows, even to the least of its elements, limitlessly in our esteem. For the man who sees nothing at the end of the world, nothing higher than himself, daily life can only be filled with pettiness and boredom. So much fruitless effort, so many wasted moments! But to those who see the synthesis of the Spirit continuing on earth beyond their own brief existence, every act and event is charged with interest and promise. Indeed, it does not matter what we do each day, or what we undergo, provided we keep a steady hand on the tiller—for are we not steering toward the fulfillment of the World? In the New Time there is no longer any distinction between those things that we classified on other levels as physical or moral, natural or artificial, organic or collective, biological or juridical. All things are seen to be supremely physical, supremely natural, supremely organic and supremely vital—according to how far they contribute to the construction and closing of the time-space cone above us.

d Fourthly, the world glows with a new warmth: that is to say, it opens wholly to the power of Love. To love is to discover and complete one's self in someone other than oneself, an act impossible of general realization on earth so long as each man can see in his neighbor no more than a closed fragment following its own course through the world. It is precisely this state of isolation that will end if we begin to discover in each other not merely the elements of one and the same thing, but of a single Spirit in search of itself. Then the medium will be established in which a basic affinity may be born and grow, springing from one seed of thought to the next, canalizing in a single direction the swarm of individual trajectories. In the old Time and Space a universal attraction of souls was inconceivable. The existence of such a power becomes possible, even inevitable, in the curvature of a world capable of noogenesis.

e Fifthly, and lastly, the Universe is illumined from within: that is to say, it shows itself to be capable of fulfilling the highest of our mystical aspirations. By virtue of the convergence of the cosmic lines, as I have said, we must surmise the existence of a higher center of consciousness ahead of us, at the apogee of Evolution. But if we seek to determine the position and analyze the properties of this Supreme Center it soon becomes clear that we must look far beyond and far above any mere aggregation of perfected Mankind. If it is to be capable of joining together in itself the prolonged fibers of the world, the apex of the cone within which we move can be conceived only as something that is ultraconscious, ultrapersonalized, ultrapresent. It must reach and act upon us, not only indirectly, through the universal network of physical synthesis, but also, and even more, directly, from center to center (that is to say, from consciousness to consciousness) by touching the most sensitive point in ourselves.

Thus it is that our humanity, renewed in its love of living and spurred on in its aspirations by the discovery that there is a peak to the arrow-course of Time, comes logically to perfect itself in an attitude of self-abandonment and adoration.

2. Toward a Christian Renewal

AS RECENTLY AS yesterday Christianity represented the highest point attained by the consciousness of Mankind in its striving to humanize itself. But does it still hold this position, or at the best can it continue to hold it for long? Many people think not; and to account for this slackening impulse in the highest and most complete of human mystical beliefs they argue that the evangelical flowering is ill-adapted to the critical and materialist climate of the modern world. They hold that the time of Christianity is past, and that some other shoot must grow in the field of religion to take its place.

But if, as I maintain, the event that characterizes our epoch is a growing awareness of the convergent nature of Space-Time, then nothing can be more ill-founded than this pessimism. Transferred within the cone of Time, and there transmuted, the Christian system is neither disorganized nor deformed. On the contrary, sustained by the new environment, it more than ever develops its main lines, acquiring an added coherence and clarity.

This is what, in conclusion, I wish to show.

What is finally the most revolutionary and fruitful aspect of our present age is the relationship it has brought to light between Matter and Spirit: spirit being no longer independent of matter, or in opposition to it,[2] but laboriously emerging from it under the attraction of God by way of synthesis and centration.

But what is the effect, for Christian faith and mysticism, of this redefinition of the Spirit? It is simply to confer absolute reality and absolute urgency upon the double dogma on which the whole of Christianity rests, and by which it is summed up: the physical primacy of Christ and the moral primacy of Charity.

Let us see.

a *The Primacy of Christ.* In the narrow, partitioned and static Cosmos wherein our fathers believed themselves to dwell, Christ was "lived" and loved by His followers, as He is today, as the Being on whom all things depend and in whom the Universe finds its "consistence." But this Christological function was not easily defended on rational grounds, at least if the attempt was made to interpret it in a full, organic sense. Accordingly Christian thinking did not especially seek to incorporate it in any precise cosmic order. At that time the Kingship of Christ could be readily expressed in terms of juridical ascendancy; or else it was sufficient

---

[2] Provided, of course, that we do not understand "matter" in a "reduplicative" and restricted sense to mean that portion of the Universe which "redescends," escaping the rising stream of Noogenesis.

that He should prevail in the nonexperimental, extracosmic sphere of the supernatural. Theology, in short, did not seem to realize that *every* kind of Universe might not be "compossible" with the idea of an Incarnation. But with the concept of Space-Time, as we have defined it, there comes into effect a harmonious and fruitful conjunction between the two spheres of rational experience and of faith. In a Universe of "Conical" structure Christ has a place (the apex!) ready for Him to fill, when His Spirit can radiate through all the centuries and all beings; and because of the genetic links running through all the levels of Time and Space between the elements of a convergent world, the Christ-influence, far from being restricted to the mysterious zones of "grace," spreads and penetrates throughout the entire mass of Nature in movement. In such a world Christ cannot sanctify the Spirit without (as the Greek Fathers intuitively perceived) uplifting and saving the totality of Matter. Christ becomes truly universal to the full extent of Christian needs, and in conformity with the deepest aspirations of our age the Cross becomes the Symbol, the Way, the very Act of progress.

b *The Primacy of Charity.* What the modern mind finds disconcerting in Christian charity is its negative or at least static aspect, and also the "detached" quality of this great virtue. "Love one another . . ." Hitherto the gospel precept has seemed simply to mean, "Do not harm one another," or, "Seek with all possible care and devotion to diminish injustice, heal wounds and soften enmities in the world around you." Hitherto, also, the "supernatural" gift of ourselves which we were required to make to God and to our neighbor appeared to be something opposed to and destructive of the bonds of feeling attaching us to the things of this world.

But if Charity is transplanted into the cone of Time nothing remains of these apparent limitations and restrictions. Within a

Universe of convergent structure the only possible way in which an element can draw closer to its neighboring elements is by *tightening the cone*—that is to say, by causing the whole layer of the world of which it is a part to move toward the apex. In such an order of things no man can love his neighbor without drawing nearer to God—and, of course, reciprocally (but this we knew already). But it is also impossible (this is newer to us) to love either God or our neighbor without assisting the progress, in its physical entirety, of the terrestrial synthesis of the spirit: since it is precisely the progress of this synthesis which enables us to draw closer together among ourselves, while at the same time it raises us toward God. Because we love, and in order that we may love even more, we find ourselves happily and especially compelled to participate in all the endeavors, all the anxieties, all the aspirations and also all the affections of the earth—*in so far as these embody a principle of ascension and synthesis.*

Christian detachment subsists wholly in this wider attitude of mind: but instead of "leaving behind" it leads on; instead of cutting off, it raises. It is no longer a breakaway but a way through; no longer a withdrawal but an act of emerging. Without ceasing to be itself, Charity spreads like an ascending force, like a common essence at the heart of all forms of human activity, whose diversity is finally synthesized in the rich totality of a single operation. Like Christ Himself, and in His image, it is universalized, it acquires a dynamic and is humanized by the fact of doing so.

To sum up, in order to match the new curve of Time Christianity is led to discover the values of this world *below the level of God*, while Humanism finds room for a God above the level of this world. Inverse and complementary movements: or rather, the two

faces of a single event which perhaps marks the beginning of a
new era for Mankind.

———— ※ ————

THIS DOUBLE TRANSFORMATION is something more than a
speculation of my own. Throughout the world at this moment,
without distinction of country, class, calling or creed, men are ap-
pearing who have begun to reason, to act and to pray in terms of
the limitless and organic dimensions of Space-Time. To the outside observer such men may still seem isolated. But they are aware
of one another among themselves, they recognize each other
whenever their paths cross. They know that tomorrow, rejecting
old concepts, divisions and forms, the whole world will see what
they see and think as they do.

PEKING, FEBRUARY 13, 1942. *PSYCHÉ*, NOVEMBER 1946.

CHAPTER 6

LIFE AND THE
PLANETS

What Is Happening at This Moment on Earth?

DURING THE FIVE years that the Earth has trembled beneath our feet, its vast human masses splitting and reforming, we have begun to be conscious of the fact that we are in the grip of forces many millions of times transcending our individual liberties. For even the most positivist and realist among us the evidence is growing that the present crisis far exceeds the economic and political factors which seemed to provoke it, and within the framework of which we may have hoped that it would remain confined. This conflict is no merely localized and temporary affair, a matter of periodical readjustment between nations. The events we are witnessing and undergoing are unquestionably bound up with the general evolution of terrestrial life; they are of *planetary dimensions*. It is therefore on the planetary scale that they must be assessed, and it is in these terms that I ask you to consider them, so that we may better understand, better endure, and, I will add, better love these things greater than ourselves which are taking place around us and sweeping us along in their course.

What does the world-adventure upon which we are embarked look like, when we seek to interpret it both objectively and hopefully in the light of the widest, soundest and most modern concepts of astronomy, geology and biology? That is what I propose to discuss here: not from the viewpoint of Sirius, as the saying is—that is to say, with the lofty detachment of an observer seeing things from so far off that they fail to touch him—but with the anxious intensity of a son of Earth who draws back in order to be able to see more deeply into the matter and spirit of a movement upon which his happiness depends.

This lecture is divided into three parts:

One.   The place of living planets in the Universe. Smallness and vastness.

Two.   The place of Man on the planet Earth—at the head.

Three. The place of our generation—our own place—in the evolution of Mankind. Assessment.

And finally a summing-up: the end of planetary life. Death or escape?

Let us begin.

I. LIVING PLANETS IN THE UNIVERSE

1. From the Point of View of the Immense: The Apparent Insignificance of the Earth

FROM WHAT WE now know of astronomy the planets would seem at first sight to be a perfectly insignificant and negligible element in the universe as a whole. How does the sidereal universe look to the eyes of modern science? No doubt you have gazed up at the sky on a fine winter's night and, like innumerable human beings before you, had an impression of a serene and tranquil firmament twinkling with a profusion of small, friendly lights, all apparently at the same distance from yourself. But telescopic and spectroscopic observation, and increasingly exact calculations, are transforming this comfortable spectacle into a vision that is very much more unsettling, one which in all probability will profoundly affect our moral outlook and religious beliefs when it has passed from the minds of a few initiates into the mass-consciousness of Mankind as a whole: immensities of distance and size, huge extremes of temperature, torrents of energy. . . .

That we may better understand what the earth means, we must try to penetrate, step by step, within this "infinity."

First, the stars.

The stars constitute the natural sidereal unit. It is toward them therefore, the analysis of their structure and the study of their distribution, that the researches of astrophysics are principally directed. The process of research is one based entirely on the analysis of light, calling for miracles of patience, ability and acumen; but it is astonishingly fruitful, since it enables exact measurements to be made of the mass, energy, diameter, distance and movement of objects vast in themselves but ultramicroscopic to us because of their remoteness.

The first thing to note is that, in certain aspects, the stars seem to vary a great deal among themselves. Certain of them, the "red giants," are of colossal dimensions, their diameter exceeding 450 times that of the Sun (if the Sun were as large as they it would extend beyond Earth, Jupiter and Saturn as far as Uranus!) Others, the "white dwarfs," are smaller than the earth; and still others, the most numerous category, closely resemble the Sun both in their dimensions and their yellow color. We find similar contrasts of brilliance and temperature. One star may be the equivalent of 300,000 suns in luminosity, whereas another may amount to only a fifty-thousandth part of it (as great a difference, the astronomer Sir James Jeans observed, as there is between a lighthouse and a glowworm). These, of course, are extreme cases. In the matter of surface temperature, if the Sun and the majority of stars are round about 6,000° Centigrade (three times the temperature of an electric arc) there are some of 11,000° (Sirius) and even of 23,000°; and on the other hand there are some as low as 3,500° (the red giants).

But beneath this great diversity, which is due principally to the varying ages of the stars, there is concealed a sort of deep identity. Whether giants, medium-sized or dwarfs, the stars are curiously similar in mass (from one to ten times the mass of the sun), which proves, incidentally, that they must vary prodigiously in their mean density—1.4 in the case of the Sun, but 50,000 and even 300,000 in the case of the dwarf stars (a fragment the size of a pinch of snuff, brought from one of these to Earth, would weigh a ton!)

So we have approximate identity of mass, and therefore calibration. If we now consider the number of the stars ($15,000 \times 10^6$ visible to the optical telescope alone) you will understand how it is possible to say, cosmically speaking, that we are enveloped in a sort of monstrous gas formed of molecules as heavy as the Sun moving at distances from each other so great that they have to be reckoned in light-years (bearing in mind that light travels at a speed of 186,000 miles per second, and that we are only eight light-minutes distant from the Sun)—*a gas made of stars!*

A gas of stars. The very conjunction of the two words is startling. But the shock is even greater when we learn that these myriads of suns scattered in the void are no more than the grains forming a supergrain of infinitely greater magnitude, and that this in its turn is no more than one unit amid a myriad of similar units! Imagination is confounded. . . . Yet this is what we learn, beyond any possibility of doubt, from the Milky Way and the other galaxies.

You will all have gazed in curiosity at the Milky Way, that long whitish ribbon which, extending from east to west over the two hemispheres, girdles our firmament. Astronomers have long felt that this mysterious, luminous train must constitute one of the most important structural features of the Universe. They sought, therefore, to decipher it, and they have succeeded in doing so. This is the conclusion, dumbfounding but certain, at which they have arrived. The Milky Way, they tell us, is not at all, as one might suppose, a sort of cloud of diffused matter drifting like a mist among the stars. Instead, it denotes the boundary, it marks the equatorial contour, of a prodigious lenticular accumulation of cosmic matter nursing, in its spiral arms, the solar system, all our constellations, all our visible stars, and further millions besides (perhaps $100{,}000 \times 10^{6}$ altogether); these latter being so remote from us that their total effect is to convey no more than a vague, milky impression to our eyes. It has been possible to calculate the dimensions of this extraordinary celestial formation and the speed with which it rotates upon itself. According to Jeans its diameter is about 200,000 light-years and it takes 3 million years to complete a single revolution, at a peripheral speed of several hundred miles per second. Compared with this stupendous disc, Jeans remarks, the Earth's orbit is no bigger than a pin's head compared with the surface of the American continent.

But the Milky Way, our Milky Way, is not the only one of its kind in the Universe. Here and there small milky patches are to be discerned in the sky, which the telescope shows to be spiral clouds containing sparks of brilliance. These, as we now know, are infinitely farther away from us than the stars. They do not belong to our own, immediate world—or, as one might put it, to the sidereal vessel which bears us. They are other islands, other fragments of the Universe, other Milky Ways sailing in convoy with our own through space (or even diverging from it at fantastic speeds). Several millions of these galaxies have already been counted (each, we must remember, composed of millions upon millions of stars), separated from one another by an average distance of 2 million light-years, and all of approximately the same size! A gas of galaxies on top of a gas of stars. . . . This is the truly overwhelming spectacle, far beyond our power to picture it, in which our present vision of the Universe culminates when we look in the direction of the Immense.

But must we not assume, following the logic of this principle of recurrence, that even beyond this there are supergalaxies, each formed of a group of spiral nebulae? We cannot be sure, but it seems improbable. The Universe is not composed, as Pascal thought, of pieces enclosed one in another, repeating themselves indefinitely and identically from bottom to top, from the infinitesimal to the immense. At a certain level the cosmic structure stops dead, and we pass on to "something else." Beyond the galaxies there is nothing, according to Einsteinian physics, unless it be the spherical frame of Space-Time within which all things move in a circle, without ever coming to an end or being able to leave it. . . . Let us put aside this still unresolved problem of the upper limits of the world, and since we do not yet know what may be beyond or around the galaxies, let us at least consider what unites them—that is to say, try to describe the genesis of their swarm. It is along this path, as you will see, that we shall eventually encounter the *planets* in search of which we started out.

At the very beginning, so the astronomers tell us—that is to say, billions of years ago—there was in place of the present world a diffused atmosphere, billions of times less dense than air, spreading in all directions over billions of miles. This "primordial chaos," as Jeans calls it, must have seemed homogeneous; but inasmuch as it was subject to the force of gravity it was excessively unstable. A slight unevenness of distribution occurring by chance at any given point (a contingency that was bound to arise) was all that was needed to cause the entire edifice to break up into parts which, sundering themselves from their neighbors, coiled in more and more tightly upon themselves in enormous clots—their vastness, by the law of celestial mechanics, being directly proportionate to the lightness of the matter of which they were originally composed. This was the first stage of the birth of the galaxies. The same disruptive process then operated within the separate galaxies, engendering smaller clots, since cosmic matter had become heavier. Thus the stars appeared.

Are we then to suppose that a third stage occurred in which the stars, in their turn, gave birth to planets through the condensation of their substance? This was the famous theory of Laplace; but a more thorough analysis of the problem has shown that it could not have happened in this way. Astronomers are today agreed that the distribution and movement of the heavenly bodies composing the solar system can only be explained by the hypothesis of a purely fortuitous occurrence—for example, the near contact of two stars. This is to say that Mercury, Venus, Earth, Mars, Jupiter, Saturn, Uranus or little Pluto, the farthest of all, would not exist had not another sun, by an extraordinary chance, passed so near to our sun as almost to touch it (within three diameters!) wresting from it, by force of attraction, a long, cigar-shaped filament which in the course of time broke up into a string of separate globes.[1]

And this brings us to the heart of the problem we set out to solve, namely: "What is the place, the significance and the importance of our planets in the Universe?"

Because of their very small dimensions (even Jupiter is a dwarf

---

[1] There is a tendency nowadays to abandon "catastrophic" theories in favor of "evolutionary" theories (a return to the Kant-Laplace nebula under a new form, cf. Weizsächer's theory). [Ed.].

compared with the Sun), the extreme weakness of the energy they radiate, and the short time they have been in existence (the galaxies were billions of years old when the solar system was born); even more important, because of their mode of existence, the planets look not merely like poor relations but like strangers and intruders in the sidereal system. Created by chance, they have no place in the normal and orthodox evolution of astral matter; with the exasperating result that we know nothing for certain about the existence or frequency of occurrence of planets outside the solar system. In Laplace's thesis almost every star should have its girdle of planets. In present-day theory perhaps one star in 100,000 (Jeans's estimate: Eddington puts the figure at millions) possesses them. And if to this we add the fact that, in the case of any given planet, it calls for a further extremely rare accident to produce the conditions which would endow it with life, we can see what a fantastically small figure, quantitatively speaking, our Earth cuts in the Universe.

I said just now, in seeking to describe the magnitude of the human events which are overtaking us, that they were of "planetary" importance. But is not "planetary" almost synonymous with "infinitesimal"? Let me recall from memory the hard words of Jeans (he wrote more hopeful ones later, you will be relieved to learn):

"What does life amount to? We have tumbled, as though through error, into a universe which by all the evidence was not intended for us. We cling to a fragment of a grain of sand until such time as the chill of death shall return us to primal matter. We strut for a tiny moment upon a tiny stage, well knowing that all our aspirations are doomed to ultimate failure and that everything we have achieved will perish with our race, leaving the Universe as though we had never existed. . . . The Universe is indifferent and even hostile to every kind of life."

But let us boldly state it: this bleak vista is not only so discour-

aging as to make action impossible; it is so much at variance, physically, with the existence and exercise of our intelligence (which, after all, is the one force in the world capable of dominating the
world) that it cannot be the last word of Science. Following the
physicists and astronomers we have thus far been contemplating
the Universe in terms of the Immense—immensity of space, time,
energy and number. But is it not possible that we have been looking through the wrong end of the telescope, or seeing things in the
wrong light? Suppose, instead, we survey the same landscape—
without, of course, attempting in any way to alter its arrangement—in its biochemical aspect, that of *Complexity*.

2. In Terms of Complexity; or the Planets as Vital Centers of the Universe

WE WILL DEFINE the "complexity" of a thing, if you allow, as the
quality the thing possesses of being composed—

a of a larger number of elements, which are
b more tightly organized among themselves.

In this sense an atom is more complex than an electron, a molecule more complex than an atom, and a living cell more complex
than the highest chemical nuclei of which it is composed, the difference depending (on this I insist) not only on the number and diversity of the elements included in each case, but at least as much
on the number and correlative variety of the links formed between
these elements. It is not, therefore, a matter of *simple* multiplicity
but of organized multiplicity; not simple complication but *centered*
complication.

This idea of complexity (more exactly, centrocomplexity) is easily grasped. In a universe where science ends by analyzing everything and taking everything apart, it simply expresses a particular characteristic applicable to every kind of body, like its mass, volume or any other dimension. But what do we gain by using this characteristic, rather than another, for the purpose of classifying the objects around us?

I will cite two advantages, although it means somewhat anticipating the latter parts of this lecture.

First, in the multitude of things comprising the world, an examination of their degree of complexity enables us to distinguish and separate those which may be called "true natural units," the ones that really matter, from the accidental pseudo-units, which are unimportant. The atom, the molecule, the cell and the living being are true units because they are both formed and centered, whereas a drop of water, a heap of sand, the Earth, the Sun, the stars in general, whatever the multiplicity or elaborateness of their structure, seem to possess no organization, no "centricity." However imposing their extent they are false units, aggregates arranged more or less in order of density.

Secondly, the coefficient of complexity further enables us to establish, among the natural units which it has helped us to "identify" and isolate, a system of classification that is no less natural and universal. Let us try to depict this classification in schematic form, as it might be drawn on a blackboard.

At the very bottom of the board we have the ninty-two simple chemical elements (from hydrogen to uranium) formed by groups of atomic nuclei together with their electrons.

Above these come the molecules composed of groups of atoms. These molecules, in the case of the carbon compounds, may become enormous. In the albuminoids (or proteins) there may be thousands of associated atoms: the molecular weight of hemoglobin is 68,000.

Above these again come the mysterious viruses, strange bodies producing a variety of maladies in animals and plants, concerning which we do not yet know if they are monstrous chemical molecules or living infrabacteria. Their molecular weight runs into millions.

Higher still we come to the lowest cells. I do not know if any attempt has yet been made to ascertain the atomic content of these (it must amount to billions) but they are undoubtedly groups of proteins.

And finally we reach the world of higher living forms, each composed of groups of cells. To take a very simple instance, that of the plant duckweed; its content is estimated to be 4×10^{20} atoms.

For the present we will disregard an even higher category which may conceivably have its place at the head of the list—that formed by the grouping, not merely of cells, but of metazoa synthetically associated in such a manner as to comprise, when taken together, a single, living superorganism. We shall come back to this.

This scheme of classification, based essentially on the intimate structure of beings, is undeniably *natural* in principle. But it can also be seen to possess a double and extreme significance.

It is significant, in the first place, because for the scientist it bridges the long-standing, troublesome and seemingly irreducible gap between biology and physics. The wide distinction, which for philosophical reasons it has been thought necessary to draw between life and matter, ceases to be valid as *a law of recurrence* comes to light, in the phenomenal field, experientially linking these two orders of phenomena. Beyond the millionth atom everything happens as though the material particles quickened and were vitalized; the Universe organizes itself in a single, grand progression, somewhat untidy no doubt, but on the whole clear in its orientation, ascending from the most rudimentary atom to the highest form of living things.

Secondly it is significant because, arranged according to our scale of complexity, the elements succeed one another *in the historical order of their birth.* The place in the scale occupied by each particle situates the element chronologically in the genesis of the Universe, that is to say, in Time. It dates it.

Thus the rising scale conforms both to the ascending movement toward higher consciousness and to the unfolding of evolutionary time. Does not this suggest that, by using the degree of complexity as a guide, we may advance very much more surely than by following any other lead as we seek to penetrate to the truth of the world and to assess, in terms of absolute values, the relative importance, the place, of all things?

With this in mind let us look again at the vast sidereal units (galaxies and suns) and this time try to assess their importance not in terms of their immensity or even complexity (since, as I have said, nebulae and stars are no more than aggregates) but in terms of the complexity of the elements which compose them.

We now see a very different picture; a complete reversal of values and perspective.

Let us look first at what is largest, the galaxies. In their least condensed parts (that is to say, in what they still contain of the vestiges of primordial chaos), the matter composing them is extremely tenuous; probably hydrogen, the most primitive substance known to us in the field of distinguishable matter. One nucleus and one electron: the simplest combination imaginable.

Now come down a stage in the scale of the immensities and look at the stars. Here the chemism is more elaborate. Whether in the red giants, the medium yellows or the white dwarfs, we may surmise the presence in the center of heavy and extremely unstable elements possessing a greater atomic weight than uranium (unless these are simply "ordinary matter" reduced to a physical state of extraordinary compression). At the same time, in the lighter surface-zone enveloping these depths the spectroscope can discern the entire range of our simple elements. In the stars, therefore, if we compare them with the original galaxies, the degree of complexity rises rapidly; but, and this is of major importance, it cannot go beyond a certain stage; that is to say (if we except a number of simple groups perceptible in the incandescent atmosphere of certain stars) it cannot reach the level of the composite bodies, i.e., the large molecules. The fact is that even on the periphery of these prodigious centers of energy the temperature is far too great for any higher combination to possess stability. The stars are essentially laboratories in which Nature, starting with primordial hydrogen, manufactures atoms. For the operation to go beyond this point we have to imagine two astonishing things:

First, that by a sort of "skimming" process a portion of the stellar substance separates from the rest, deriving entirely from the surface-zone of lighter atoms which are not constantly threatened with radioactive disintegration. The larger molecules can only be constructed of elements possessing almost unlimited stability.

Secondly, that this light and stable "cream" of any given star, having escaped beyond the reach of the tempest of energy blazing at the heart of the parent-body, may yet remain sufficiently close to it to derive a moderate benefit from its radiations: for the large molecules need energy for their synthesis.

But are not these two providential occurrences (the selection of a suitable "dough" and its treatment in a suitable "oven") precisely what that mysterious body, our father-star, effected in a single operation when, passing close to our Sun, it detached from its surface and scattered over a wide distance the ribbon of matter that became the planets?

You will now see where my argument is tending, or more exactly, where the guide which we have elected to follow, the scale of complexity, is irresistibly leading us. Despite their vastness and splendor the stars cannot carry the evolution of matter much beyond the atomic series: it is only on the very humble planets, on them alone, that the mysterious ascent of the world into the sphere of high complexity has a chance to take place. However inconsiderable they may be in the history of sidereal bodies, however accidental their coming into existence, the planets are finally nothing less than the key-points of the Universe. It is through them that the axis of Life now passes; it is upon them that the energies of an Evolution principally concerned with the building of large molecules is now concentrated.

We may well be dismayed by the rarity and improbability of heavenly bodies such as our own. But does not everyday experience teach us that in every order of Nature, and at every level, nothing succeeds except at the cost of prodigious waste and fantastic hazards? A monstrously fragile conjunction of chances normally dictates the birth of the most precious and essential beings. We can only bow before this universal law whereby, so strangely to our minds, the play of large numbers is mingled and confounded with a final purpose. Without being overawed by the Improbable, let us now concentrate our attention on the planet we call Earth. Enveloped in the blue mist of oxygen which its life breathes, it floats at exactly the right distance from the sun to enable the higher chemisms to take place on its surface. We do well to look at it with emotion. Tiny and isolated though it is, it bears clinging to its flanks the destiny and future of the Universe.

II. MAN ON THE PLANET EARTH:
THE MOST COMPLEX OF MOLECULES

HAVING ESTABLISHED, ON the basis of complexity, the astral preeminence of the planets in the sidereal system, and particularly that of the Earth, our obvious next step is to seek to determine, in cosmic terms, the significance and value on Earth of what we very improperly call "the human species."

If the essential function and dignity of the Earth consist in its being one of the rare laboratories where, in time and space, the synthesis of ever larger molecules is proceeding; and if, as our table of complexity shows, living organisms, far from originating in germs fallen upon Earth from the celestial spaces, are simply the highest composites to spring from planetary geochemism,[2] then the discovery of Man's absolute place in the Universe becomes simply a matter of deciding what position we who constitute Mankind occupy in the evolving range of supermolecules.

Here, however, a difficulty arises.

Where relatively simple molecular units are concerned their order of complexity may be roughly expressed by the number of atoms they contain, their "corpuscular number" as one might call it. But when this corpuscular number exceeds a million (from the virus on) and still more when we come to the higher forms of life (there are something like a hundred billion cells in an average mammal, and hundreds of millions of atoms to each cell!) it becomes impossible to estimate the number of atoms, which would be so vast as to be almost meaningless even if it could be calculated. At this level of organization, in fact, the actual number of atoms contained in complex units is of minor importance compared with the number and quality of the *links* established between the atoms.

---

[2] I need hardly point out that for the purpose of this lecture, which does not seek to go outside the field of scientific observation, only the succession and interdependence of phenomena are taken into account: that is to say, *an experimental law of recurrence,* not an ontological analysis of causes.

HOW THEN ARE we to go about classifying the higher living units
so that the position of Man in terms of complexity may be determined? What method shall we adopt?

We can do it very simply by introducing what is called a
change of variable. The more complex a being is, so our Scale of
Complexity tells us, the more is it centered upon itself and therefore the more conscious does it become. In other words, the higher
the degree of complexity in a living creature, the higher its consciousness; and vice versa. The two properties vary in parallel and
simultaneously. If we depict them in diagrammatic form, they are
equivalent and interchangeable. So it comes to this, that when we
have reached the point where complexity can no longer be reckoned in number of atoms we can nevertheless continue to measure
it (and accurately) by noting the increase of consciousness in the
living creature—in practical terms, the development of its nervous
system. This is the solution of our problem.

Accordingly, if we use the factor of psychic growth (or, which
comes to the same thing, progress of cerebralization) as a scale
whereby we may measure the growth of complexity through the
maze of invertebrates, arthropods and vertebrates, the position
and significance of the human type in nature at once becomes apparent. For of all the numberless types of living units that have appeared in the course of the last 300 million years, Man, judged by
his power of reflection (itself bound up with the ultracomplexity of
a brain composed of many millions of cells) not only comes indisputably first, but occupies a place of his own at the head of all the
other "very great complexes" evolved on Earth. And this incidentally explains why he tends increasingly to break away from the rest
of terrestrial life, to detach himself in such a manner as to form
(we shall return to this) a separate planetary envelope.

What does this mean except that, having been led by the idea of complexity to consider the Earth one of the vital points of the Universe, we find ourselves compelled, following the same principle, to recognize in Man the most advanced, and therefore the most valuable, of all the planetary elements? If it is the Earth which bears the fortunes of the world, then it is Man, in his extreme centro-complexity, who bears the fortunes of the Earth.

But if that is our situation, what is our destiny?

III. THE PRESENT STATE OF MANKIND:
THE PHASE OF PLANETIZATION

TO OPEN ANY book treating scientifically, philosophically or sociologically of the future of the Earth (whether by a Bergson or a Jeans) is to be struck at once by a presupposition common to most of their authors, certain biologists excepted. Explicitly or by inference they talk as though Man today had reached a final and supreme state of humanity beyond which he cannot advance; or, in the language of this lecture, that, Matter having attained in *Homo sapiens* its maximum of centro-complexity on Earth, the process of super-moleculization on the planet has for good and all come to a stop.

Nothing could be more depressing, but also, fortunately, more arbitrary and even scientifically false, than this doctrine of immobility. No proof exists that Man has come to the end of his potentialities, that he has reached his highest point. On the contrary, everything suggests that at the present time we are entering a peculiarly critical phase of superhumanization. This is what I hope to persuade you of by drawing your attention to an altogether extraordinary and highly suggestive condition of the world around us, one which we all see and are subject to, but without paying any attention to it, or at least without understanding it: I mean the increasingly rapid growth in the human world of the forces of collectivization.

The phenomenon calls for no detailed description. It takes the form of the all-encompassing ascent of the masses; the constant tightening of economic bonds; the spread of financial and intellectual associations; the totalization of political regimes; the closer physical contact of individuals as well as of nations; the increasing impossibility of being or acting or thinking *alone*—in short, the rise, in every form, of the *Other* around us. We are all constantly aware of these tentacles of a social condition that is rapidly evolving to the point of becoming monstrous. You feel them as I do, and probably you also resent them. If I were to ask your views you would doubtless reply that, menaced by this unleashing of blind forces, there is nothing we can do but evade them to the best of our ability, or else submit, since we are the victims of a sort of natural catastrophe against which we are powerless and in which there is no meaning to be discerned.

But is it true that there is nothing to understand? Let us look more closely, once again by the light of our principle of complexity.

The first thing to give us pause, as we survey the progress of human collectivization, is what I would call the inexorable nature of a phenomenon which arises directly and automatically out of the conjunction of two factors, both of a structural kind: first, the confined surface of the globe, and secondly, the incessant multiplication, within this restricted space, of human units endowed by ever-improving means of communication with a rapidly increasing scope for action; to which may be added the fact that their advanced psychic development makes them preeminently capable of influencing and interpenetrating one another. Under the combined effect of these two natural pressures a sort of mass-concretion of Mankind upon itself comes of necessity into operation.

But, the second noteworthy point, the phenomenon of concretion, or cementing, turns out to be no sudden or unpredictable event. Looking at the picture as a whole we see that Life, from its lowest level, has never been able to effect its syntheses except through the progressively closer association of its elements, whether in the oceans or on land. Upon an imaginary earth of constantly increasing extent, living organisms, being only loosely associated, might well remain at the monocellular stage (if indeed they got so far); and certainly Man, if free to live in a scattered state, would never have reached even the neolithic stage and social development. The totalization in progress in the modern world is in fact nothing but the natural climax and paroxysm of a process of grouping which is fundamental to the elaboration of organized matter. Matter does not vitalize or supervitalize itself except by compression.

I do not think it is possible to reflect upon this twofold inrooting, both structural and evolutionary, which characterizes the social events affecting us, without being at first led to the surmise, and finally overwhelmed by the evidence, that the collectivization of the human race, at present accelerated, is nothing other than a higher form adopted by the process of moleculization on the surface of our planet. The first phase was the formation of proteins up to the stage of the cell. In the second phase individual cellular complexes were formed, up to and including Man. We are now at the beginning of a third phase, the formation of an organicosocial supercomplex, which, as may easily be demonstrated, *can only occur* in the case of *reflective, personalized elements*. First the vitalization of matter, associated with the grouping of molecules; then the hominization of Life, associated with a supergrouping of cells; and finally the *planetization* of Mankind, associated with a *closed* grouping of people: Mankind, born on this planet and spread over its entire surface, coming gradually to form around its earthly matrix a single, major organic unity, enclosed upon itself; a single, hypercomplex, hypercentered, hyperconscious arch-molecule, coextensive with the heavenly body on which it was born. Is not this what is happening at the present time—the closing of this spherical, thinking circuit?

This idea of the planetary totalization of human consciousness (with its unavoidable corollary, that wherever there are life-bearing planets in the Universe, they too will become encompassed, like the Earth, with some form of planetized spirit) may at first sight seem fantastic: but does it not exactly correspond to the facts, and does it not logically extend the cosmic curve of moleculization? It may seem absurd, but in its very fantasy does it not heighten our vision of Life to the level of other and universally accepted fantasies, those of atomic physics and astronomy? However mad it may seem, the fact remains that great modern biologists, such as Julian Huxley and J. B. S. Haldane, are beginning to talk of Mankind, and to predict its future, as though they were dealing (all things being equal) with a brain of brains.

So why not?

Clearly this is a matter in which I cannot compel your assent. But I can assure you, of my own experience, that the acceptance of this organic and realistic view of the social phenomenon is both eminently satisfying to our reason and fortifying to our will.

Satisfying to the intelligence above all. For if it be true that at this moment Mankind is embarking upon what I have called its "phase of planetization," then everything is clarified, everything in our field of vision acquires a new sharpness of outline.

The tightening network of economic and psychic bonds in which we live and from which we suffer, the growing compulsion to act, to produce, to think collectively which so disquiets us—what do they become, seen in this way, except the first portents of the superorganism which, woven of the threads of individual men, is preparing (theory and fact are at one on this point) not to mecha-

nize and submerge us, but to raise us, by way of increasing complexity, to a higher awareness of our own personality?

The increasing degree, intangible, and too little noted, in which
present-day thought and activity are influenced by the passion for
discovery; the progressive replacement of the workshop by the laboratory, of production by research, of the desire for well-being by
the desire for *more*-being—what do these things betoken if not the
growth in our souls of a great impulse toward superevolution?

The profound cleavage in every kind of social group (families,
countries, professions, creeds) which during the past century has
become manifest in the form of two increasingly distinct and irreconcilable human types, those who believe in progress and those
who do not—what does this portend except the separation and
birth of a new stratum in the biosphere?

Finally, the present war; a war which for the first time in history is as widespread as the earth itself; a conflict in which human
masses as great as continents clash together; a catastrophe in which
we seem to be swept off our feet as individuals—what aspect can
it wear to our awakened eyes except that of a crisis of birth, almost
disproportionately small in relation to the vastness of what it is destined to bring forth?

Enlightenment, therefore, for our intelligence. And, let it be
added, *sustenance and necessary reassurance for our power of will.* Through
the centuries life has become an increasingly heavy burden for Man
the Species, just as it does for Man the Individual as the years pass.
The modern world, with its prodigious growth of complexity, weighs
incomparably more heavily upon the shoulders of our generation
than did the ancient world upon the shoulders of our forebears. Have
you never felt that this added load needs to be compensated for by an
added passion, a new sense of purpose? To my mind, this is what is
"providentially" arising to sustain our courage—the hope, the belief
that some immense fulfillment lies ahead of us.

If Mankind were destined to achieve its apotheosis, if Evolution were to reach its highest point, in our small, separate lives, then indeed the enormous travail of terrestrial organization into which we are born would be no more than a tragic irrelevance. We should all be dupes. We should do better in that case to stop, or at least to call a halt, destroy the machines, close the laboratories, and seek whatever way of escape we can find in pure pleasure or pure nirvana.

But if on the contrary Man sees a new door opening above him, a new stage for his development; if each of us can believe that he is working so that the Universe may be raised, in him and through him, to a higher level—then a new spring of energy will well forth in the heart of Earth's workers. The whole great human organism, overcoming a momentary hesitation, will draw its breath and press on with strength renewed.

Indeed, the idea, the hope of the planetization of life is very much more than a mere matter of biological speculation. It is more of a necessity for our age than the discovery, which we so ardently pursue, of new sources of energy. It is this idea which can and must bring us the spiritual fire without which all material fires, so laboriously lighted, will presently die down on the surface of the thinking earth: the fire inspiring us with the joy of action and the love of life.

All this, you may say to me, sounds splendid: but is there not another side to the picture? You tell us that this new phase of human evolution will bring about an extension and deepening of terrestrial consciousness. But do not the facts contradict your argument? What is actually happening in the world today? Can we really detect any heightening of human consciousness even in the most highly collectivized nations? Does it not appear, on the contrary, that social totalization leads directly to spiritual retrogression and greater materialism?

My answer is that I do not think we are yet in a position to judge recent totalitarian experiments fairly: that is to say, to decide whether, all things considered, they have produced a greater degree of enslavement or a higher level of spiritual energy. It is too early to say. But I believe this can be said, that in so far as these first attempts may seem to be tending dangerously toward the subhuman state of the ant hill or the termitary, it is not the principle of totalization that is at fault but the clumsy and incomplete way in which it has been applied.

We have to take into account what is required by the law of complexity if Mankind is to achieve spiritual growth through collectivization. The first essential is that the human units involved in the process shall draw closer together, not merely under the pressure of *external* forces, or solely by the performance of material acts, but directly, center to center, through *internal* attraction. Not through coercion, or enslavement to a common task, but through *unanimity* in a common spirit. The construction of molecules ensues through atomic affinity. Similarly, on a higher level, it is through *sympathy*, and this alone, that the human elements in a personalized universe may hope to rise to the level of a higher synthesis.

It is a matter of common experience that within restricted groups (the pair, the team) unity, far from diminishing the individual, enhances, enriches and liberates him in terms of himself. True union, the union of heart and spirit, does not enslave, nor does it neutralize the individuals which it brings together. It *superpersonalizes* them. Let us try to picture the phenomenon on a terrestrial scale. Imagine men awakening at last, under the influence of the ever-tightening planetary embrace, to a sense of universal solidarity based on their profound community, evolutionary in its nature and purpose. The nightmares of brutalization and mechanization which are conjured up to terrify us and prevent our advance are at once dispelled. It is not harshness or hatred but a new kind of love, not yet experienced by man, which we must learn to look for as it is borne to us on the rising tide of planetization.

Reflecting, even briefly, on the state of affairs which might evoke this universal love in the human heart, a love so often vainly dreamed of, but which now leaves the fields of Utopia to reveal itself as both possible and necessary, we are brought to the following conclusion: that for men upon earth, all the earth, to learn to love one another, it is not enough that they should know themselves to be members of one and the same *thing*; in "planetizing" themselves they must acquire the consciousness, without losing themselves, of becoming one and the same *person*. For (and this is writ large in the Gospel) there is no total love that does not proceed from, and exist within, that which is personal.

And what does this mean except, finally, that the planetization of Mankind, if it is to come properly into effect, presupposes, in addition to the enclosing Earth, and to the organization and condensation of human thought, yet another factor? I mean the rise on our inward horizon of a cosmic *spiritual* center, a supreme pole of consciousness, upon which all the separate consciousnesses of the world may converge and within which they may love one another: the *rise of a God*.

It is here that reason may discern, conforming to and in harmony with the law of complexity,[3] an acceptable way of envisaging "the end of the world."

IV. THE END OF PLANETARY LIFE:
MATURITY AND WITHDRAWAL

THE END OF the world—for us, that is to say, the end of Earth. . . . Have you ever thought seriously, *in human terms*, about that somber and certain eventuality?

---

[3] Which here culminates, we may note, in a sort of proof of the existence of God: "proof by complexity."

Life at the beginning seemed modest in its requirements. A few hours in the sun were all it seemed to ask and all it needed to justify itself in its own eyes. But this was only a semblance, belied at the earliest stages of vitalization by the tenacity with which the most humble cells reproduce themselves and multiply. This tenacity continues through all the enormous effusion of the animal kingdom, and bursts into the light of day with the appearance, in thinking Man, of the formidable power of prevision. It cannot but grow still more imperious with every forward stride of human consciousness. I have spoken of the impulse to act, without which there can be no action. But in practice it is not enough, if the impulse is to be sustained in face of the ever-growing onslaughts of the *taedium vitae*, for it to be offered nothing more than an immediate objective, even though this be as great as the planetization of Mankind. We must strive for ever more greatness; but we cannot do so if we are faced by the prospect of an eventual decline, a disaster at the end. With the germ of consciousness hatched upon its surface, the Earth, our perishable earth threatened by the final, absolute zero, has brought into the Universe a demand, henceforth irrepressible, not only that all things shall not die, but that what is best in the world, that which has become most complex, most highly centered, shall be saved. It is through human consciousness, genetically linked to a heavenly body whose days are ultimately numbered, that Evolution proclaims its challenge: either it must be *irreversible*, or it need not go on at all! Man the individual consoles himself for his passing with the thought of the offspring or the works which he leaves behind. But what will presently be left of Mankind?

Thus every attempt to situate Man and the Earth in the framework of the Universe comes inevitably upon the heavy problem of death, not of the individual but on the planetary scale—a death which, if we seriously contemplate it, must instantly paralyze all the vital forces of the Earth.

In an attempt to dispel this shadow Jeans calculated that the Earth has many millions of years of habitability ahead of it, so that Man is still only on the threshold of his existence. He bade us warm our hearts, in this fresh dawn, with the *almost* limitless prospects of the glorious day that is only beginning. But a few pages previously he had talked of Mankind sadly growing old and disillusioned on a chilling globe, faced by inevitable extinction. Does not that first thought destroy the second?

Others seek to reassure us with the notion of an escape through space. We may perhaps move to Venus—perhaps even further afield. But apart from the fact that Venus is probably not habitable (is there water?) and that, if journeying between celestial bodies were practicable, it is hard to see why we ourselves have not already been invaded, this does no more than postpone the end.

We cannot resolve this contradiction, between the congenital mortality of the planets and the demand for irreversibility developed by planetized life on their surface, by covering it up or deferring it: we have finally to banish the specter of Death from our horizon.

And this we are enabled to do by the idea (a corollary, as we have seen, of the mechanism of planetization) that ahead of, or rather in the heart of, a universe prolonged along its axis of complexity, there exists a divine center of convergence. That nothing may be prejudged, and in order to stress its synthesizing and personalizing function, let us call it the *point Omega*. Let us suppose that from this universal center, this Omega point, there constantly emanate radiations hitherto only perceptible to those persons whom we call "mystics." Let us further imagine that, as the sensibility or response to mysticism of the human race increases with planetization, the awareness of Omega becomes so widespread as to warm the earth psychically while physically it is growing cold. Is it not conceivable that Mankind, at the end of its totalization, its folding-in upon itself, may reach a critical level of maturity where, leaving Earth and stars to lapse slowly back into the dwindling mass of primordial energy, it will detach itself from this planet and join the one true, irreversible essence of things, the Omega point? A phenomenon perhaps outwardly akin to death: but in reality a simple metamorphosis and arrival at the supreme synthesis. An escape from the planet, not in space or outwardly, but spiritually and inwardly, such as the hypercentration of cosmic matter upon itself allows.

This hypothesis of a final maturing and ecstasy of Mankind, the logical conclusion of the theory of complexity, may seem even more farfetched than the idea (of which it is the extension) of the planetization of Life. Yet it holds its ground and grows stronger upon reflection. It is in harmony with the growing importance which leading thinkers of all denominations are beginning to attach to the phenomenon of mysticism. In any event, of all the theories which we may evolve concerning the end of the Earth, it is the only one which affords a coherent prospect wherein, in the remote future, the deepest and most powerful currents of human consciousness may converge and culminate: intelligence and action, learning and religion.

LECTURE DELIVERED AT THE FRENCH EMBASSY IN PEKING,

MARCH 10, 1945. *ÉTUDES*, MAY 1946.

CHAPTER 7

A GREAT EVENT
FORESHADOWED: THE
PLANETIZATION OF MANKIND

Argument

UNDERLYING ALL THE surface changes of present-day history, the reality and paramount importance of a single basic event is becoming daily more manifest: namely, the rise of the masses, with its natural corollary, the socialization of Mankind. The supreme interest and significance of this process lies in the fact that, scientifically analyzed, it may be seen to be irresistible in two ways: in the *planetary* sense, because it is associated with the closed shape of the earth, the mechanics of generation and the psychic properties of human matter; and in the *cosmic* sense because it is the expression and prolongation of the primordial process whereby, at the uttermost extreme from the disintegrating atom, psychic force is born into the Universe and continuously grows, fostered by the ever more complicated grouping of matter. Projected forward, this law of recurrence makes it possible for us to envisage a future state of the Earth in which human consciousness, reaching the climax of its evolution, will have attained a maximum of complexity, and, as a result, of concentration by total "reflexion" (or *planetization*) of itself upon itself.

Although our individualistic instincts may rebel against this drive toward the collective, they do so in vain and wrongly. In vain, because no power in the world can enable us to escape from what is in itself the power of the world. And wrongly because the real nature of this impulse that is sweeping us toward a state of super-organization is such as to make us more completely personalized and human.

The very fact of our becoming aware of this profound ordering of things will enable human collectivization to pass beyond the *enforced* phase, where it now is, into the *free* phase: that in which (men having at last understood that they are inseparably joined elements of a converging Whole, and having learnt in consequence to *love* the preordained forces that unite them) a natural union of affinity and sympathy will supersede the forces of compulsion.

Preamble

IT HAS BECOME very difficult, in the world's present state of upheaval and distraction, to form any idea of the significance of what is going on except by rising above the individual level. So many opposing forces (ideas, passions, institutions, peoples) meet and clash around us that to the thinking person it may well appear that the human ship is rudderless in the storm. Are we going ahead or astern, or are we simply hove-to? No means of telling while we remain at sea level: the waves hide the horizon.

I can see only one way of escape from this state of uncertainty which threatens to paralyze all positive action: we must rise above the storm, the chaos of surface detail, and from a higher vantage-point look for the outline of some great and significant phenomenon. To rise up so as to see clearly is what I have tried to do, and it has led me to accept, however improbable they may appear, the reality and the consequences of the major cosmic process which, for want of a better name, I have called "human planetization."

Despite appearances and a certain overlapping due to the vastness of the subject (as we draw near to the Whole, physics, metaphysics and religion strangely converge) I am prepared to maintain that what I have to say does not anywhere go beyond the field of scientific observation. What this essay claims to offer is not philosophical speculation but an extension of our biological perspective—no more, and no less.

1. An Irresistible Physical Process: *The Collectivization of Mankind*

WE MIGHT SUPPOSE, if we set out to examine the state of things on the morrow of the most fearful convulsion that has ever shaken the living strata of the Earth, that we should find the soil mined and fissured to its depths. So great a shock must surely have exposed all the points of weakness, unloosed all the forces of dispersal and divergence and left Mankind shattered within itself. This is what we might expect to find.

But instead of this state of universal ruin, and if we disregard the psychological haze of fatigue and resentment which, as I shall show, is only a passing phase, what do we actually see?

Geographically, since 1939 a vast expanse of the earth, the region of the Pacific, hitherto on the fringe of civilization, has for practical purposes entered irrevocably into the orbit of industrialized nations. Mechanized masses of men have invaded the south-

ern seas, and up-to-date airfields have been permanently installed
on what were until yesterday the poetically lost islands of Polyne-
sia.

Ethnically, during the same space of time, there has been a
vast and pitiless confusion of peoples, whole armies being removed
from one hemisphere to the other, and tens of thousands of
refugees being scattered across the world like seed borne on the
wind. Brutal and harsh though the circumstances have been, who
can fail to perceive the inevitable consequences of this new stirring
of the human dough?

And finally, during the same period, economically and psychically the entire mass of Mankind, under the inexorable pressure of
events and owing to the prodigious growth and speeding up of the
means of communication, has found itself seized in the mould of
a communal existence—large sections tightly encased in countless
international organizations, the most ambitious the world has ever
known; and the whole anxiously involved in the same passionate
upheavals, the same problems, the same daily news. . . . Can anyone seriously suppose that we shall be able to rid ourselves of
habits such as these?

No; during these six years, despite the unleashing of so much
hatred, the human block has not disintegrated. On the contrary, in
its most rigid organic depths it has further increased its viselike grip
upon us all. First 1914–1918, then 1939–1945—two successive turns
of the screw. Every new war, embarked upon by the nations for the
purpose of detaching themselves from one another, merely results
in their being bound and mingled together in a more inextricable
knot. The more we seek to thrust each other away, the more do we
interpenetrate.

Indeed, how could it be otherwise?

Confined within the geometrically restricted surface of the
globe, which is steadily reduced as their own radius of activity in-

creases, the human particles do not merely multiply in numbers every day at an increasing rate, but through contact with one another automatically develop around themselves an ever denser tangle of economic and social relationships. Moreover, being each exposed at the core of their being to the countless spiritual influences emanating from the thought, the will and the passions of all their fellow creatures they find themselves constantly subjected in spirit to an enforced rule of resonance. It must surely be clear that, under the pressure of these relentless factors—relentless because they are a part of the deepest and most generalized conditions of the planetary structure—there is only one way in which the tide can flow: the way of ever-increasing unification. In speculating on the earthly destiny of Man we are accustomed to say that in the ultimate future nothing is certain, except that a day must come when our planet will be uninhabitable. But for those who are not afraid to look ahead, another thing awaits us that is no less certain. As the Earth grows older, so does its living skin contract, and ever more rapidly. The last day of Man will coincide for Mankind with the maximum of its tightening and in-folding upon itself.

I know that to see determinisms everywhere in history may be to oversimplify and is certainly dangerous. Every so often authoritative voices are raised protesting that there is no fateful significance in the rise of the masses, or the planned economy or the growth of democracy. Where details and modalities are concerned, these defenders of individual liberty are often right. But they go astray, or will do so, if in their proper spirit of opposition to everything that is passive and blind in the world they seek to close their eyes, and ours, to the overriding superdeterminism which irresistibly impels Mankind to converge upon itself.

Whether we like it or not, from the beginning of our history and through all the interconnected forces of Matter and Spirit, the process of our collectivization has ceaselessly continued, slowly or in jerks, gaining ground each day. That is the fact of the matter. It is as impossible for Mankind not to unite upon itself as it is for the human intelligence not to go on indefinitely deepening its thought! . . . Instead of seeking, against all the evidence, to deny or disparage the reality of this grand phenomenon, we do better to accept it frankly. Let us look it in the face and see whether, using it as an unassailable foundation, we cannot erect upon it a hopeful edifice of joy and liberation.

2. The One Possible Interpretation:
A Superorganization of the Matter Around Us

TO UNDERSTAND THE significance of the world forces of collectivization, and what it is that they so imperiously demand of us, we need to look down from a great height and contemplate, in their widest, most general aspect, the organic relationships linking *consciousness* and *complexity* within the Universe.

It would seem that Man, observing it with curiosity, has always been aware of the law of compensation whereby, in every circumstance of nature, the most highly spiritualized souls are associated with the most corruptible and intricate bodies. But it has remained for modern biology and biochemistry to disclose this contrast, which lay observation could do no more than perceive, in all its persistence and sharpness. We marvel, in the light of recent developments of microscopic and chemical analysis, at the formidable edifice of atoms and varied mechanisms which are found to exist in living creatures, *the more living they are.* How has it happened that, faced by this constant balance between physical plurality and psychic unity, we have been so slow to grasp the possibility of a physical link of causality connecting them? Hints of the existence of such a link are today beginning to crop up everywhere in scientific works. Let me venture, in a schematic and personal way, to interpret this line of inquiry which, explicitly or by implication, is gradually attracting the notice of philosophers and scientists.

Before doing anything else we must dismiss from our picture of the world the factitious barrier which, to ordinary perception, separates the so-called inanimate particles (atoms, molecules, etc.) from living particles or bodies. That is to say, we must assume, on the strength of their common *behavior* (multiplicity in similarity) that all, in their varied degrees of complexity and magnitude, are manifestations of a single, fundamental, granular structural principle of the Universe—simply larger or smaller particles.

And having done this let us postulate in principle that consciousness (like such phenomena as the variation of mass according to speed, or radiation in relation to temperature) is a universal property common to all the particles constituting the Universe, but varying in proportion to the complexity of any particular molecule: which amounts to saying that the degree of psychism, the "within," of the different elements composing the world will be small or great, according to the place of the element in the astronomically extended scale of complexities at present known to us.

The effect of this double modification is to transform our perception of things. Hitherto, in the eyes of a Science too much accustomed to constructing the world on one spatial axis extending in a line from the infinitely small to the infinitely great, the larger molecules of organic chemistry, and still more the living cellular composites, have existed without any defined position, like wandering stars, in the general scheme of cosmic elements. Now however, simply by the introduction of another dimension, a new order and definition become apparent. Traversing the rising axis from the infinitesimal to the immense another branch appears, rising through Time from the infinitely simple to the supremely complicated. It is on this branch that the consciousness-phenomenon has its place and eventually shows itself. There is first a long, obscure stretch which seems dead but is in fact "imperceptibly alive." Then, at the stage of particles reaching a million atoms in their complexity (viruses), we come to the first flush heralding the dawn of Life. Later, after the cell, there is a definite radiation growing richer and more intense with the formation and gradual concentration of nervous systems. And finally, at the extreme end of the known spectrum, comes the thinking incandescence of the human brain.

By this reordering of things, not only does Life, despite its extreme rarity and localization in Space, show itself, in symmetry with atomic disintegration, to be a fundamental universal current (*the current*); not only does Man, with his billions of interacting nervous cells, find a natural, cosmically enrooted place in this generalized physical scheme; but something begins to take shape ahead of Man. Once again we find ourselves confronted by the forces of collectivization.

Owing to an inhibition, inherent in our mentality, which prevents us from looking squarely at collectivity, "common sense" has long refused to accept any but superficial analogies between the "moral or artificial" sphere of human institutions and the "physical" sphere of organized Nature. Indeed, it is only very recently, and as yet timidly, that sociology has ventured to set up the first bridges between biology and itself. But once we have accepted the general Law of Recurrence linking the growth of consciousness to the advance of complexity within a process of universal evolution, nothing can arrest the logical sequence in which two worlds which we were accustomed to regard as being completely separate are seen to approach and complement each other. We see Nature combining molecules and cells in the living body to construct separate individuals, and the same Nature, stubbornly pursuing the same course but on a higher level, combining individuals in social organisms to obtain a higher order of psychic results. The processes of chemistry and biology are continued without a break in the social sphere. This accounts for the tendency, which has been insufficiently noted, of every living phylum (insect and vertebrate) to group itself toward its latter end in socialized communities. Above all, in the case of Man (the only living species in which the variety, quality and intensity of individual relationships enables the phenomenon to achieve its full extent) it explains the rapid psychic rise accompanying socialization, which takes the following forms:

a the appearance of a collective memory in which a common inheritance of Mankind is amassed in the form of accumulated experience and passed on through education;

b the development, through the increasingly rapid transmission of thought, of what is in effect a generalized nervous system, emanating from certain defined centers and covering the entire surface of the globe;

c the emergence, through the interaction and ever-increasing concentration of individual viewpoints, of a faculty of common vision penetrating beyond the continuous and static world of popular conception into a fantastic but still manageable world of atomized energy.

All around us, tangibly and materially, the thinking envelope of the Earth—the Noosphere—is adding to its internal fibers and tightening its network; and at the same time its internal temperature is rising, and with this its psychic potential. These two associated portents allow of no misunderstanding. What is really going on, under cover and in the form of human collectivization, is the superorganization of Matter upon itself, which as it continues to advance produces its habitual, specific effect, the further liberation of consciousness. It is all one and the same process. And, by very reason of the elements involved, the process cannot achieve stability until, over the entire globe, the human *quantum* has not merely closed the circle upon itself (as it is doing at this moment, in a penultimate phase) but has become organically *totalized*.

So what finally lies ahead of us is a planetary arrangement of human mass and energy, coinciding with a maximal radiation of thought—at once the external and internal "planetization" of Mankind. That is what we are inexorably heading for, in the tightening embrace of the social determinisms. The Earth could more easily evade the pressures which cause it to contract upon itself, the stars more readily escape from the spatial curve which holds them on their headlong courses, than we men can resist the cosmic forces of a converging universe!

And why should we seek to resist these unifying forces which are essentially benevolent? Is it because we are afraid that in the process of supercreation they will render us less human?

The basic characteristic of Man, the root of all his perfections, is his gift of consciousness *in the second degree*. Man not only knows; he knows that he knows. He reflects. But this power of reflection, when restricted to the individual, is only partial and rudimentary. As Nietzsche has rightly observed, although he put the wrong construction on it, the individual, faced by himself alone, cannot know himself exhaustively. It is only when opposed to other men that he can discover his own depth and wholeness. However personal and incommunicable it may be at its root and origin, Reflection can only be developed in communion with others. It is essentially a *social* phenomenon. What can this mean except that its eventual completion and wholeness must exactly coincide (in full accord with the Law of Complexity) with what we have called the plane-tization of Mankind?

Some hundreds of thousands of years ago Consciousness achieved the stage of its *own centration*, and thus the power of thought, in a brain that had reached the limit of nervous compli-cation: this was the first stage in the hominization of Life on earth.

In due course, after the passage of further thousands or even millions of years, it can, and it must, supercenter itself in the bosom of a Mankind *totally reflexive upon itself.*

Instead of vainly opposing or meekly submitting to the cre-ative forces of the planet which bears us, should we not rather let our lives be illumined and broadened in the growing light of this second stage of hominization?

3. The Only Permissible Inward Reaction: The Spirit of Evolution

A REMARKABLE CHANGE overtakes the process of zoological evolution at the level of Man. Until that point was reached every animal, feebly separated from its fellows, existed largely for the purpose of preserving and developing its own species, so that for the individual life was primarily a matter of propagation. But from the time of Man a sort of internal granulation seems to attack the Tree of Life, causing it to disintegrate at the top. With the dawn-ing of Reflection each conscious unit isolates itself and, one would say, tends increasingly to live only for itself, as though, by the fact of hominization, the phylum were broken up into individuals; and as though, in the hominized individual, the *phyletic sense* were sub-merged until it finally vanishes.

It is to this alarming course of psychic decomposition, and at the very moment when it seems to be reaching its crisis, that the prospect of a human planetary fulfillment brings the appropriate remedy. If, as we have shown, the social phenomenon is not merely a blind determinism but the portent, the inception of a second phase of human *Reflexion* (this time not merely individual but collective), then it must mean that the phylum is reconstituting itself above our heads in a new form, a new ramification, no longer of divergence but of convergence; and consequently it is the *Spirit of Evolution* which, suppressing the spirit of egoism, is of its own right springing to new life in our hearts, and in such a way as to counteract those elements in the forces of collectivization which are poisonous to Life.

That the construction of superorganisms is a hazardous operation (like all Life's major transformations) is something of which we find ample evidence in the study of animal colonies, or, where Man is concerned, the spectacle afforded by recent totalitarian experiments. We are alarmed by all forms of communized existence, and not without reason, because they seem automatically to entail the loss or mutilation of our individual personality. But may it not be that our fear of a process of mechanization seemingly fatal to our activities arises simply from the fact that we have left the most important element out of our reckoning? In the foregoing paragraphs I have deliberately, for the sake of objectivity, looked only at the external or enforced aspect of human planetization. Thus far we have taken no account of the internal reactions to be expected of planetized matter. But what happens if we consider the "planetizing" process as applied not merely to a passive substratum but to a human mass inspired with the Spirit of Evolution? What we then see is a flood of *sympathetic* forces, spreading from the heart of the system, which transforms the whole nature of the phenomenon: sympathy in the first place (an act of quasi-adoration) on the part of all the elements gathered together for the general impulse that carries them along; and also the sympathy (this time fraternal) of each separate element for all that is most unique and incommunicable in each of the coelements with which it converges in the unity, not only of a single act of vision but of a single living subject. But to say "love" is to say "liberty." There need be no fear of enslavement or atrophy in a world so richly charged with charity!

Therefore, *provided it be accompanied by a revival of the phyletic sense*, the collectivization of the Earth shows itself to be the true instrument, not merely of cerebral superhominization but of complete humanization. By interiorizing itself under the influence of the Spirit of Evolution, planetization (as the theory of complexity would lead us to expect) can physically have but one effect: it can only personalize us more and more, and eventually (as can be demonstrated by following to their conclusion all the successive stages of its twofold demand for wholeness and irreversibility) "divinize" us through access to some Supreme Center of universal convergence.

But the question arises, will this universal Spirit of Evolution (the necessary antidote and natural reaction to the growth of complexity in a world that has reached the stage of Reflexion) come when it is needed? Will it flower in time to ensure that, arrived at the point of superhumanity, we avoid dehumanizing ourselves? Theory may predict its imminent appearance: but have we in fact specific reasons for believing that it will truly awaken at the expected moment in the hearts of our fellow men?

4. *Deeper Than Our Present Discords: Mankind in the Reshaping*

ALTHOUGH IN TERMS of its biological, economic and mental determinisms the human earth, emerging from war, may be seen to be more tightly fastened upon itself than ever before, in its other and freer aspects it may give a first impression of growing disorder.

As I said at the beginning, a thick fog of confusion and dissension is at present drifting over the world. Indeed one might say that men have never more vehemently rebuffed and detested one another than they do now, when everything drives them closer together. Is this state of moral chaos really to be reconciled with the idea and the hope that we are advancing toward *unanimity* through the closer contact of our bodies and minds?

Let us look at things more closely to see whether, even in those troubled regions of the heart, there may not be gleams of light heralding the planetization of Mankind.

Traced in broad outline the "psychic" map of the world would show on its surface a mosaic of vertically separated compartments (ethnical, political, religious) whereas in depth a loosening surface, symbolizing class-antagonism, would separate the human mass into two thicknesses over its entire planetary extent. Such is the tangled skein which the war has inevitably thrown into relief. Along these ancient or recent lines of division, the tightening up of the world could not fail to introduce play into the structure of the Noosphere and cause it to burst open. But what effect has it had in younger and more elastic zones?

A new substance has recently appeared in the heart of the thinking "magma"—a new element, not yet catalogued but of supreme importance: We might call it *Homo progressivus*, that is to say, the man to whom the terrestrial future matters more than the present. A new type of man indeed, when we consider that, less than two hundred years ago, the notion of an organic evolution of the World in Time had acquired neither form nor substance in the human mind. When we come to look for them, men of this sort are easily recognizable. They are scientists, thinkers, airmen and so on—all those possessed by the demon (or the angel) of Research. Let us try to plot their statistical distribution on our imaginary map. The diagram turns out to have some remarkable features.

In the first place, points denoting this new human type will be found to be scattered more or less all over the thinking face of the globe. Although more numerous among the white peoples, and as one goes lower down the social scale, they will appear, at least occasionally, in every compartment into which the human race is divided. Their emergence is clearly related to some phenomenon of a noospheric kind.

Secondly, some apparent attraction draws these scattered elements together and causes them to unite among themselves. You have only to take two men, in any gathering, endowed with this mysterious sense of the future. They will gravitate instinctively toward one another in the crowd; they will recognize one another.

But the third characteristic, the most noteworthy of all, is that this meeting and grouping together is not confined to individuals belonging to the same category or having the same origins, that is to say, belonging to the same compartment within the Noosphere. No racial, social or religious barrier seems to be effective against this force of attraction. I myself have experienced this a hundred times, and anyone who chooses can do the same. Regardless of the country, creed or social status of the person I approach, provided the same flame of expectation burns in us both, there is a profound, definitive and total contact instantly established between us. It matters nothing that differences of education or training cause us to express our hopes in different ways. We feel that we are of the same kind, and we find that our very differences are a common armor, as though there were a dimension of life in which all striving makes for nearness, not only within a corporate body but heart to heart.

I believe that these various characteristics can be accounted for in only one way. We have to accept that, accelerated by the successive intellectual and social upheavals that have shaken the world during the past century and a half, a radical process of differentiation and segregation is taking place within the human mass. And it is following precisely the course we would expect: the spontaneous individualization and separation of that which moves and rises from that which remains immobile; the irresistible multiplication and aggregation, over the whole extent of the globe, of elements activated by a (hominized) reawakening of the phyletic sense; the gradual formation and emergence, at variance with most former categories, of a new noospheric zone in which human collectivization, hitherto enforced, is at last entering its *sympathetic phase* under the influence of the newly manifest Spirit of Evolution.

It would seem, then, that the grand phenomenon which we are now witnessing represents a new and possibly final division of Mankind, based no longer on wealth but on belief in progress.

The old Marxist conflict between producers and exploiters becomes out-dated—at the best a misplaced approximation. What finally divides the men of today into two camps is not class but an attitude of mind—the spirit of movement. On the one hand there are those who simply wish to make the world a comfortable dwelling place; on the other hand, those who can only conceive of it as a machine for progress—or better, an organism that is progressing. On the one hand the "bourgeois spirit" in its essence, and on the other the true "toilers of the Earth," those of whom we may safely predict that, without violence or hatred, simply by biological predominance, they will tomorrow constitute the human race. On the one hand the castoffs; on the other, the agents and elements of planetization.

PEKING, DECEMBER 25, 1945. *CAHIERS DU MONDE NOUVEAU,*
AUGUST–SEPTEMBER 1946.

CHAPTER 8

SOME REFLECTIONS ON THE SPIRITUAL REPERCUSSIONS OF THE ATOM BOMB

ONE EARLY DAWN in the "bad lands" of Arizona, something over a year ago, a dazzling flash of light, strangely brilliant in quality, illumined the most distant peaks, eclipsing the first rays of the rising sun. There followed a prodigious burst of sound. . . . The thing had happened. For the first time on earth an atomic fire had burned for the space of a second, industriously kindled by the science of Man.

But having thus realized his dream of creating a new thunderclap, Man, stunned by his success, looked inward and sought by the glare of the lightning his own hand had loosed to understand its effect upon himself. His body was safe; but what had happened to his soul?

I shall not seek to discuss or defend the essential morality of this act of releasing atomic energy. There were those, on the morrow of the Arizona experiment, who had the temerity to assert that the physicists, having brought their researches to a successful conclusion, should have suppressed and de-

stroyed the dangerous fruits of their invention. As though it were
not every man's duty to pursue the creative forces of knowledge
and action to their uttermost end! As though, in any event, there
exists any force on earth capable of restraining human thought
from following any course upon which it has embarked!

Neither shall I here attempt to examine the economic and political problems created by the intrusion of nuclear energy upon
human affairs. How is the use of this terrifying power to be organized and controlled? This is for the worldly technicians to answer. It is sufficient for me to recall the general condition which is
necessary for the solution of the problem: it must be posed on an
international scale. As the American journal, *The New Yorker*, observed with remarkable penetration on August 18, 1945: "Political
plans for the new world, as shaped by statesmen, are not fantastic
enough. The only conceivable way to catch up with atomic energy
is with political energy directed to a universal structure."

The aim of these reflections—more narrowly concerned with
our separate souls, but for that reason perhaps going deeper—is
simply to examine, in the case of the atomic bomb, the effects of
the invention upon the inventor, arising out of the fact of the invention. Each of our actions, and the more so the more novel the
action, has its deep-seated repercussions upon our subsequent inner orientation. To fly, to beget, to kill for the first time—these, as
we know, suffice to transform a life. By the liberation of atomic energy on a massive scale, and for the first time, man has not only
changed the face of the earth; he has by the very act set in motion
at the heart of his being a long chain of reactions which, in the brief
flash of an explosion of matter, has made of him, virtually at least,
a new being hitherto unknown to himself.

Let me try, in a first approximation, to distinguish the main
links in this chain.

a At that crucial instant when the explosion was about to happen (or not happen) the first artificers of the atom bomb were crouched on the soil of the desert. When they got to their feet after it was over, it was Mankind who stood up with them, instilled with a *new* sense of power. Certainly the power was of a kind which Man had many times felt emanating from himself, in great pulsations, during the course of his history. He had felt it, for example, in the darkness of the palaeolithic age when for the first time he ventured to put fire to his own use, or accidentally discovered how to produce it; in neolithic times when he found that by cultivating thin ears of grass he could turn them into rice and millet and corn; and much later, at the dawn of our industrial era, when he found that he could tame and harness not only wild steeds but the tireless energies of steam and electricity. Each of these new conquests signified extensively and intensively, for Man and for the earth, a total rearrangement of life, a change of epoch; but when all is said they did not bring about any essential *change of plane* in the depths of human consciousness. For in all these cases (even the most beneficial, that of electricity), what did the discovery lead to except the control and utilization of forces already at liberty in the surrounding world? They called for ingenuity and adaptiveness rather than any act of creation; they were no more, in each case, than a new sail hoisted to catch a new wind. But the discovery and liberation of atomic power bears quite another aspect and in consequence has had a very different effect upon Man's soul. Here it is no longer a question of laying hands upon existing forces freely available for his use. This time a door has been decidedly forced open, giving access to a new and supposedly inviolable compartment of the universe. Hitherto Man was using matter to serve his needs. Now he has succeeded in seizing and manipulating the sources commanding the very origins of matter—springs so deep that he can release for his own purposes what seemed to be the exclusive property of the sidereal powers, and so powerful that he must think twice before committing some act which might destroy the earth. In the glow of this triumph how can he feel otherwise than exalted as he has never been since his birth; the more so since the prodigious event is not the mere accidental product of a futureless chance but the long-prepared outcome of intelligently concerted action?

b Therefore, a new sense of power: but even more, the sense of a power capable of development *to an indefinite extent*. What gripped the throats of those bold experimenters in Arizona, in that minute before the explosion, must surely have been far less the thought of the destruction it might lead to than of the critical test which the pyramid of calculation and hypothesis culminating in this solemn moment was about to undergo. The quicker ending of the war, the vast sums of money spent—what did such things matter when the very worth of science itself was on trial? That vast and subtle edifice of equations, experiments, interwoven calculations put together little by little in the laboratories, would it survive the test of this culminating experiment which would make of it, in everyday terms, something tangible, efficacious, unanswerable? Was it a dream or reality? This was the moment of truth. In a few instants they would know. . . .

And the flame truly sprang upward at the place and time prescribed, energy did indeed burst forth from what, to ordinary perception, was inert, noninflammable matter. Man at that moment found himself endowed not merely with his existing strength but with a method which would enable him to master all the forces surrounding him. For one thing he had acquired absolute and final confidence in the instrument of mathematical analysis which for the past century he had been forging. Not only could matter be expressed in terms of mathematics, it could be subjugated by mathematics. Perhaps even more important, he had discovered, in the unconsidered unanimity of the act which circumstances had forced upon him, another secret pointing the way to his omnipotence. For the first time in history, through the nonfortuitous conjunction of a world crisis and an unprecedented advance in means of communication, a planned scientific experiment employing units of a hundred or a thousand men had been successfully completed. And very swiftly. In three years a technical achievement had been realized which might not have been accomplished in a century of isolated efforts. Thus the greatest of Man's scientific triumphs happens also to be the one in which the largest number of brains were enabled to join together in a single organism, at the same time more complex and more centered, for the purpose of research. Was this simply coincidence? Did it not rather show that in this as in other fields nothing in the universe can resist the converging energies of a sufficient number of minds sufficiently grouped and organized?

Thus considered, the fact of the release of nuclear energy, overwhelming and intoxicating though it was, began to seem less tremendous. Was it not simply the first act, even a mere prelude, in a series of fantastic events which, having afforded us access to the heart of the atom, would lead us on to overthrow, one by one, the many other strongholds which science is already besieging? The vitalization of matter by the creation of supermolecules. The remodeling of the human organism by means of hormones. Control of heredity and sex by the manipulation of genes and chromosomes. The readjustment and internal liberation of our souls by direct action upon springs gradually brought to light by psychoanalysis. The arousing and harnessing of the unfathomable intellectual and effective powers still latent in the human mass. . . . Is not every kind of effect produced by a suitable arrangement of matter? And have we not reason to hope that in the end we shall be able to arrange every kind of matter, following the results we have obtained in the nuclear field?

c It is thus, step by step, that Man, pursuing the flight of his growing aspirations, taught by a first success to be conscious of his power, finds himself impelled to look beyond any purely mechanical improvement of the earth's surface and increase of his external riches, and to dwell upon the *growth and biological perfection of himself.* A vast accumulation of historical research and imaginative reconstruction already existed to teach him this. For millions of years a tide of knowledge has risen ceaselessly about him through the stuff of the cosmos; and that in him which he calls his "I" is nothing other than this tide atomically turning inward upon itself. This he knew already; but without knowing to what extent he could render effective aid to the flood of life pouring through him. But now, after that famous sunrise in Arizona, he can no longer doubt. He not only can but, of organic necessity, he *must* for the future assist in his own genesis. The first phase was the creation of mind through the obscure, instinctive play of vital forces. The second phase is the rebounding and acceleration of the upward movement through the reflexive play of mind itself, the only principle in the world capable of combining and using for the purpose of Life, *and on the planetary scale,* the still-dispersed or slumbering energies of matter and of thought. It is broadly in these terms that we are obliged henceforth to envisage the grand scheme of things of which, by the fact of our existence, we find ourselves a part.

So that today there exists in each of us a man whose mind has been opened to the meaning, the responsibility and the aspirations of his cosmic function in the universe; a man, that is to say, who whether he likes it or not has been transformed into another man, in his very depths.

   d   The great enemy of the modern world, "Public Enemy No. 1," is *boredom*. So long as Life did not think, and above all did not have *time* to think—that is to say, while it was still developing and absorbed with the immediate struggle to maintain itself and advance—during all that time it was untroubled by questions as to the value and interest of action. Only when a margin of leisure for reflection came to intervene between the task and its execution did the workman experience the first pangs of *taedium vitae*. But in these days the margin is immeasurably greater, so that it fills our horizon. Thanks to the mechanical devices which we increasingly charge with the burden not only of production but also of calculation, the quantity of unused human energy is growing at a disturbing rate both within us and around us; and this phenomenon will reach its climax in the near future, when nuclear forces have been harnessed to useful work. I repeat: despite all appearances, Mankind is bored. Perhaps this is the underlying cause of all our troubles. We no longer know what to do with ourselves. Hence in social terms the disorderly turmoil of individuals pursuing conflicting and egoistical aims; and, on the national scale, the chaos of armed conflict in which, for want of a better object, the excess of accumulated energy is destructively released . . . "Idleness, mother of all vices."

   But these lowering storm clouds are what the Sense of Evolution, arising in human consciousness, is destined to disperse. Whatever may be the future economic repercussions of the atom bomb, whether over- or underestimated, the fact remains that in laying hands on the very core of matter we have disclosed to human existence a supreme purpose: the purpose of pursuing ever further, to the very end, the forces of Life. In exploding the atom we took our first bite at the fruit of the great discovery, and this was enough for a taste to enter our mouths that can never be washed away: the taste for supercreativeness. It was also enough to ensure that the nightmare of bloody combat must vanish in the light of some form of growing unanimity. We are told that, drunk with its own power, mankind is rushing to self-destruction, that it will be consumed in the fire it has so rashly lit. To me it seems that thanks to the atom bomb it is war, not mankind, that is destined to be eliminated, and for two reasons. The first, which we all know and long for, is that the very excess of destructive power placed in our hands must render all armed conflict impossible. But what is even more important, although we have thought less about it, is that war will be eliminated at its source in our hearts because, compared with the vast field for conquest which science has disclosed to us, its triumphs will soon appear trivial and outmoded. Now that a true objective is offered us, one that we can only attain by striving with all our power in a concerted effort, our future action can only be convergent, drawing us together in an atmosphere of sympathy. I repeat, sympathy, because to be ardently intent upon a common object is inevitably the beginning of love. In affording us a biological, "phyletic" outlet directed upward, the shock which threatened to destroy us will have the effect of giving us a sense of direction and a dynamic force and finally (within certain limits) of making us of one mind. The atomic age is not the age of destruction but of union in research. For all their military trappings, the recent explosions at Bikini herald the birth into the world of a Mankind both inwardly and outwardly pacified. They proclaim the coming of the *Spirit of the Earth*.

e We are at the precise point where, if we are to restore complete equilibrium to the state of psychic disarray which the atomic shock has induced in us, we must sooner or later (sooner?) decide upon our attitude to a fundamental choice; the point where our conflicts may begin again, and fiercely, but by other means and on a different plane.

I spoke of the Spirit of the Earth. What are we to understand by that ambiguous phrase?

Is it the Promethean or Faustian spirit: the spirit of autonomy and solitude; Man with his own strength and for his own sake opposing a blind and hostile Universe; the rise of consciousness concluding in an act of possession?

Is it the Christian spirit, on the contrary: the spirit of service and of giving; Man struggling like Jacob to conquer and attain a supreme center of consciousness which calls to him; the evolution of the earth ending in an act of union?

Spirit of force or spirit of love? Where shall we place true heroism, where look for true greatness, where recognize objective truth?

It would take too long, and it is outside the scope of this paper, to discuss the comparative worth of two opposed forms of adoration, the first of which may well have attracted poets, but only the second of which, I think, presents itself to the reflective mind as capable of conferring upon a universe in motion its full spiritual coherence, its full consistence in its passage through death, and finally its whole message for our hearts.[1]

What does matter here, on the other hand, is to note that Mankind cannot go much further along the road upon which it has embarked through its latest conquests without having to settle (or be divided intellectually on) the question of which summit it must seek to attain.

---

[1] Witnesses of that experiment in Arizona found, in the anguish of the last instants, that in the depths of their hearts they were *praying*. (Official Report: appendices.)

In short, the final effect of the light cast by the atomic fire into the spiritual depths of the earth is to illumine within them the overriding question of the ultimate end of Evolution—that is to say, the problem of God.

ÉTUDES, SEPTEMBER 1946.

CHAPTER 9

FAITH IN PEACE

I AM NO politician and still less a prophet. Like the rest of us I am anxiously following the proceedings at the Peace Conference, this poignant spectacle of the two halves of Mankind wrangling incessantly over points of detail but making no fundamental contact because they approach everything from different angles. How will it all turn out? I do not know.

But I am, if I may be allowed the term, a "geobiologist," and I have looked hard and long at the face of Mother Earth. For this reason I feel and I am convinced of one thing: that nothing is more dangerous for the future of the world, nothing moreover less warranted in Nature, than the affected resignation and false realism with which in these days a great number of people, hunching their shoulders and drawing in their heads, predict (and in so doing tend to provoke) a further catastrophe in the near future. More than all the remnants of hatred lingering between nations, this terror of inevitable war, which sees no cure for warfare except in even greater terror, is responsible for poisoning the air we breathe. That is why, humbly and devoutly echoing a divine utterance, I feel the need to cry to those around me, "What do ye fear, O men of little faith?" Do you not see that the peace which you no longer dare to hope for (when you do not actually scorn it as a myth) is possible and indeed certain, provided you will grasp what the word "peace" means and what it requires of you? Let me beg you to rise for a moment above the dust and smoke obscuring the horizon and gaze with me at the course of the world.

I

IN THE FIRST place I maintain that peace—I mean, some form of universal and stable peace—is *possible* in human terms. Why should it not be? Of course we can easily pile up arguments and evidence to refute this hopeful view. Historically, there have always been wars, and they have grown more harsh: therefore there will continue to be war till the end of time. Morally, man is evil, and becoming more so as he grows more civilized: what grounds have we for hoping that he will improve? And scientifically, since what characterizes the development of the animal species from its beginning is the struggle for life, how can we expect, mere humans that we are, to escape from this essential biological condition without which there can be neither growth nor progress? I am well aware of the many reasons for skepticism, which as a geobiologist I have pondered as much as anyone. But I must say frankly that none of them impresses me, because to my mind all are neutralized and finally annulled by a fact of higher importance to which, I do not know why, sociologists seem to pay no attention. I mean the particular and *unique* structure of the zoological group to which we belong. Until the coming of Man the branches or shoots composing the different living species tended inexorably to diverge and spread ever more widely apart as they de-

veloped. With Man, on the other hand, owing to the grand psychological phenomenon of Reflection, the branches of his species follow an entirely different course. Instead of separating and detaching
themselves from one another they turn inward and presently inter-
twine, so that by degrees, races, peoples, nations merging together,
they come to form a sort of uniconscious superorganism. To eyes
that can see, this is what is now happening. And having noted this
profound change in the evolutionary process at the human level,
how can we fail to see that it changes the whole nature of the problem, so that, in seeking to forecast the development of human society in this matter of war and peace, we cannot simply project the
history of the animal world into the future, or even that of the first
hundred or two hundred millennia of our own species? Biologically
speaking, what has hitherto driven living creatures to mutual destruction has clearly been the necessity which impelled them to supplant one another in order to survive. But why should their survival
depend upon their supplanting one another, except for the reason
that they existed independently of one another? Ultimately and fundamentally it is the divergence of the living branches, operating
from the highest level down to the family and the individuals composing the family, which has always been the cause of human
conflict. But suppose, on the contrary (this is the entirely new development in the case of the human race) that the outspreading and
unfolding of forms gradually gives way to a process of in-folding.
Then the previous economy of Nature undergoes a radical change:
for converging branches do not survive by eliminating each other;
they have to unite. Everything that formerly made for war now
makes for peace, and the zoological laws of conservation and survival must wear an opposite sign if they are to be applied to Man.
The whole phenomenon has been reversed. This may well account
for the terrible upheavals we have undergone; not an irresistible increase in the tide of war, but simply a clash of currents: the old dis-

ruptive surface forces driving against a merging in the depths which is already taking place. Why not, after all?

2

IT IS HARD to escape the conclusion, looking at things in this way, that despite all appearances to the contrary Mankind is not only capable of living in peace but by its very structure *cannot fail eventually to achieve peace.* Here, of course, we encounter the formidable element of human freedom of action, of which it is endlessly repeated that its unpredictable interference with the established proceedings of Nature threatens constantly to disrupt and frustrate them. But we need to be clear about this. No doubt it is true that up to a point we are free *as individuals* to resist the trends and demands of Life. But does this mean (it is a very different matter) that we can escape collectively from the fundamental set of the tide? I do not think so. When I consider the inexorable nature of the universal impulse which for more than six hundred million years has ceaselessly promoted the global rise of consciousness on the earth's surface, driving on through an endlessly multiplying network of opposing hazards; when I reflect upon the irresistible forces (geographical, ethnic, economic and psychic) whose combined effect is to thrust the human mass ever more tightly in upon itself; when finally, on the occasion of some great act of human collaboration or devotion, I perceive as though in a lightning flash the prodigious, still-slumbering affinity which draws the "thinking molecules" of the world together—wherever I look I am forced to the same conclusion: that the earth is more likely to stop turning than is Mankind, as a whole, likely to stop organizing and unifying itself. For if this interior movement were to stop, it is the Universe itself, embodied in Man, that would fail to curve inward and achieve totalization. And nothing, as it seems, can prevent the universe from succeeding—nothing, not even our human liberties, whose essential tendency to union may fail in detail but cannot (without "cosmic" contradiction) err "statistically." According to whether one looks at it from the point of view of the isolated unit or in terms of all units taken together, the human synthesis, that is to say, Peace, shows us two complementary faces (like so many other things in this world): first a steep slope, only to be climbed by constant effort in the face of many setbacks; and ultimately the point of balance to which the whole system must inevitably come.

3

PEACE THEREFORE IS certain: it is only a matter of time. Inevitably, with an inevitability which is nothing but the supreme expression of liberty, we are moving laboriously and self-critically toward it. But what exactly do we mean by this—*what kind of peace?* Only a peace, it is perfectly clear, which will allow, express and correspond to what I have called the vital in-folding of Mankind upon itself. A sustained state of growing convergence and concentration, a great organized endeavor: if it is not that kind of peace, then what I have been saying is worthless and we are back with our uncertainties. This means that all hope of bourgeois tranquillity, all dreams of "millenary" felicity in which we may be tempted to indulge, must be washed out, eliminated from our horizon. A perfectly ordered society with everyone living in effortless ease within a fixed framework, a world in a state of tranquil repose, all this has nothing to do with our advancing Universe, apart from the fact that it would rapidly induce a state of deadly tedium. Although, as I believe, concord must of necessity eventually prevail on earth, it can by our premises only take the form of some sort of tense cohesion pervaded and inspired with the same energies, now become harmonious, which were previously wasted in bloodshed: unanimity in search and conquest, sustained among us by the universal resolve to raise ourselves upward, all straining shoulder to shoulder, toward even greater heights of consciousness and freedom. In short, true peace, the only kind that is biologically possible, betokens neither the ending nor the reverse of warfare, but war in a naturally sublimated form. It reflects and corresponds to the normal state of Mankind become at last alive to the possibilities and demands of its evolution.

And here a last question arises, bringing us to the heart of the problem. Why is it, finally, that men at this very moment are still so painfully incapable of agreeing among themselves; why does the threat of war still appear so menacing? Is it not because they have still not purged themselves sufficiently of the demon of immobilism? Is not the underlying antagonism which separates them at the conference tables quite simply the eternal conflict between motion and inertia, the cleavage between one part of the world that moves and another that does not seek to advance? Let us not forget that faith in peace is not possible, not justifiable, except in a world dominated by faith in the future, *faith in Man* and the progress of Man. By this token, so long as we are not all of one mind, and with a sufficient degree of ardor, it will be useless for us to seek to draw together and unite. We shall only fail.

That is why, when I look for reassurance as to our future, I do not turn to official utterances, or "pacifist" manifestations, or conscientious objectors. I turn instinctively toward the ever more numerous institutions and associations of men where in the search for knowledge a new spirit is silently taking shape around us—the soul of Mankind resolved at all costs to achieve, in its total integrity, the uttermost fulfillment of its powers and its destiny.

CHAPTER 10

THE FORMATION OF
THE NOOSPHERE[1]

A Plausible Biological Interpretation of Human History

GRADUALLY, BUT BY an irresistible process (since and through the work of Auguste Comte, Cournot, Durkheim, Lévy-Bruhl and many others) the organic is tending to supersede the juridical approach in the concepts and formulations of sociol-

[1] Note in the *Revue des Questions Scientifiques* where this essay originally appeared: "To avoid misunderstanding it may be well to point out that the general synthesis outlined in these pages makes no claim to replace or to exclude the theological account of human destiny. The description of the *Noosphere* and its attendant biology, as here propounded, is no more opposed to the Divine Transcendence, to Grace, to the Incarnation or to the ultimate Parousia, than is the science of palaeontology to the Creation, or of embryology to the First Cause. The reverse is true. To those prepared to follow the author in his thinking it will be apparent that biology merges into theology, and that the Word made Flesh is to be regarded, not as a postulate of science—which would be in the nature of things absurd—but as something, a mysterious Alpha and Omega, taking its place within the whole plan of the universe, both human and divine." Pierre Charles, S.J.

ogists. A sense of collectivity, arising in our minds out of the evolutionary sense, has imposed a framework of entirely new dimensions upon all our thinking; so that Mankind has come to present
itself to our gaze less and less as a haphazard and extrinsic association of individuals, and increasingly as a biological entity
wherein, in some sort, the proceedings and the necessities of the
universe in movement are furthered and achieve their culmination.
We feel that the relation between Society and Social Organism is
no longer a matter of symbolism but must be treated in realistic
terms. But the question then arises as to how, in this shifting of values, this passage from the juridical to the organic, we may correctly
apply the analogy. How are we to escape from metaphor without
falling into the trap of establishing absurd and oversimplified par-
allels which would make of the human species no more than a kind
of composite, living animal? This is the difficulty which modern
sociology encounters.

It is with the idea and in the hope of advancing toward a solution of the problem that I here venture, basing my argument on
the widest possible zoological and biological grounds, to put forward a coherent view of the "thinking Earth" in which I believe
we may find undistorted but yet embodying the corrections required by a change of order, the whole process of Life and of vitalization.

To the natural scientist Mankind offers a profoundly enigmatic
object of study. Anatomically, as Linnaeus perceived, Man differs
so little from the other higher primates that, in strict terms of the
criteria normally applied in zoological classification, his group represents no more than a very small offshoot, certainly far less than
an Order, within the framework of the category as a whole. But in
"biospherical" terms, if I may be allowed the word, man's place on
earth is not only predominant but to a certain extent exclusive
among living creatures. The small family of hominids, the last

shoot to emerge from the main stem of Evolution, has of itself achieved a degree of expansion equal to, or even greater than, that of the greatest vertebrate layers (reptile or mammal) that ever inhabited the earth. Moreover, at the rate it is going, we can already foresee the day when it will have abolished or domesticated all other forms of animal and even plant life.

What does this mean?

I believe that the paradox will disappear and the contradictions be reconciled (with the immediate prospect of a vast field of progress for the new sociology) if we adopt the following premises:

a We must first give their place in the mechanism of biological evolution to the special forces released by the psychic phenomenon of hominization;

b Secondly we must enlarge our approach to encompass the formation, taking place before our eyes and arising out of this factor of hominization, of a particular biological entity such as has never before existed on earth—the growth, outside and above the biosphere,[2] of an added planetary layer, an envelope of thinking substance, to which, for the sake of convenience and symmetry, I have given the name of the Noosphere.[3]

Let us pursue the matter by successively examining (without at any time leaving the plane of scientific thought):

1. The birth (or, what amounts to the same thing, the zoological structure);

2. The anatomy;

[2] This term, invented by Suess, is sometimes interpreted (Vernadsky) in the sense of the "terrestrial zone containing life." I use it here to mean the actual layer of vitalized substance enveloping the earth.

[3] From *noos*, mind: the terrestrial sphere of thinking substance.

3. The physiology;

4. Finally, the principal phases of growth of the Noosphere.

1. Birth and Zoological Structure of the Noosphere

I HAVE REFERRED to the almost contradictory aspect which the
section "homo" in the order of primates assumes in the eyes of
natural scientists: that of a single family suddenly emerging, at the
end of the Tertiary era, to achieve the dimensions of a zoological
layer in itself.

If we are to appreciate this strange phenomenon we must look
back over the normal development of living forms before the coming of man. It can be characterized in two words: from its first beginnings it never ceased to be "phyletic" and "dispersive." Phyletic
in the first place: every species (or group of species) formed a sort
of shoot (or phylum) which was obliged to evolve "orthogeneti-
cally"[4] along certain prescribed lines (reduction or adaptation of
limbs, complication of teeth, increased specialization as carnivores
or herbivores, runners, burrowers, swimmers, flyers, etc.); and secondly dispersive, since the different phyla separated at certain
points of proliferation, certain "knots" which we may suppose to
be periods of particularly active mutation.[5] Until the coming of
man the pattern of the Tree of Life was always that of a fan, a
spread of morphological radiations diverging more and more,

---

[4] The word "orthogenesis" is here used in its widest sense: "A definite orientation offsetting the effect of chance in the play of heredity."

[5] Dr. A. Blanc has recently given the name of "lysis" to this phenomenon of the
releasing of morphological forces.

each radiation culminating in a new "knot" and breaking into a fan of its own.

But at the human level a radical change, seemingly due to the spiritual phenomenon of Reflection, overtook this law of development. It is generally accepted that what distinguishes man psychologically from other living creatures is the power acquired by his consciousness of turning in upon itself. The animal knows, it has been said; but only man, among animals, knows that he knows. This faculty has given birth to a host of new attributes in men— freedom of choice, foresight, the ability to plan and to construct, and many others. So much is clear to everyone. But what has perhaps not been sufficiently noted is that, still by virtue of this power of Reflection, living hominized elements become capable (indeed are under an irresistible compulsion) of drawing close to one another, of communicating, finally of uniting. The centers of consciousness, acquiring autonomy as they emerge into the sphere of reflection, tend to escape from their own phylum, which granulates into a line of individuals. Instead they pass tangentially into a field of attraction which forces one toward another, fiber to fiber, phylum to phylum: with the result that the entire system of zoological radiations which in the ordinary course would have culminated in a knot and a fanning out of new divergent lines, now tends to fold in upon itself. In time, with the reflexion of the individual upon himself, there comes an inflexion, then a clustering together of the living shoots, soon to be followed (because of the biological advantage which the group gains by its greater cohesion) by the spread of the living complex thus constituted over the whole surface of the globe. The critical point of reflexion for the biological unit becomes the critical point of "inflexion" for the phyla, which in turn becomes the point of "circumflexion" (if I may use the word) for the whole sheaf of inward-folding phyla. Or, if you prefer, the reflective coiling of the individual upon himself leads to the coiling of the phyla upon each other, which in turn leads to the coiling of the whole system about the closed convexity of the celestial body which carries us. Or we may talk in yet other terms of psychic centration, phyletic intertwining and planetary envelopment: three genetically associated occurrences which, taken together, give birth to the Noosphere.

Viewed in this aspect, entirely borne out by experience, the collective human organism which the economists so hazily envisage emerges decisively from the mists of speculation to take its place and assume the brilliance of a clearly defined star of the first magnitude in the zoological sky. Until this point was reached Nature, in her generalized effort of "complexification," to which I shall return later, had failed for lack of suitable material to achieve any grouping of individuals outside the family structure (the termitary, the ant hill, the hive). With man, thanks to the extraordinary agglutinative property of thought, she has at last been able to achieve, throughout an entire living group, a total synthesis of which the process is still clearly apparent, if we trouble to look, in the "scaled" structure of the modern human world. Anthropologists, sociologists and historians have long noted, without being very well able to account for it, the enveloping and concretionary nature of the innumerable ethnic and cultural layers whose growth, expansion and rhythmic overlapping endow humanity with its present aspect of extreme variety in unity. This "bulbary" appearance becomes instantly and luminously clear if, as suggested above, we regard the human group, in zoological terms, as simply a normal sheaf of phyla in which, owing to the emergence of a powerful field of attraction, the fundamental divergent tendency of the evolutionary radiations is overcome by a stronger force inducing them to converge. In present-day mankind, within (as I call it) the Noosphere, we are for the first time able to contemplate, at the very top of the evolutionary tree, the result that can be produced by a synthesis not merely of individuals but of entire zoological shoots.

Thus we find ourselves in the presence, in actual possession, of the superorganism we have been seeking, of whose existence we were intuitively aware. The collective mankind which the sociologists needed for the furtherance of their speculations and formulations now appears scientifically defined, manifesting itself in its proper time and place, like an object entirely new and yet awaited in the sky of life. It remains for us to observe the world by the light it sheds, which throws into astonishing relief the great ensemble of everyday phenomena with which we have always lived, without perceiving their reality, their immediacy or their vastness.

2. Anatomy of the Noosphere

IT MAY BE said, speaking in very general terms, that in asserting the zoological nature of the Noosphere we confirm the sociologists' view of human institutions as organic. Once the exceptional, but fundamentally biological, nature of the collective human complex is accepted, nothing prevents us (provided we take into account the modifications which have occurred in the dimensions in which we are working) from treating as authentic organs the diverse social organisms which have gradually evolved in the course of the history of the human race. Directly Mankind, from the nature of its origin, presents itself to our experience as a true superbody, the internal connections of this body, by reason of homogeneity, can only be treated and understood as superorgans and supermembers. Thus, for example (due allowance being made for the change of scale and environment), it becomes legitimate to talk in the sphere of economics of the existence and development of a circulatory or a nutritional system applicable to Mankind as a whole.

That we must proceed slowly and critically in this attempt to construct an "anatomy" of society is evident. Used without discernment and a profound knowledge of biology, the procedure is in danger of lapsing into puerile and sterile subtleties. But progressively pursued, and proceeding from certain major fields of knowledge, the method shows itself to be both fruitful and illuminating. This is what I shall seek to demonstrate in the three spheres of culture, machinery and research, by successively "dissecting" first the hereditary, then the mechanical and finally the cerebral apparatus of the Noosphere.

a The apparatus of heredity. One of the paradoxes attaching to the human species, a cause of some bitterness among biologists, is that every man comes into the world as defenseless, and as incapable of finding his way single-handed in our civilization, as the newborn Sinanthropus a hundred thousand years ago. As Jean Rostand[6] has remarked, during the many centuries man has striven to improve himself the fruits of his labors have brought about no organic change in him, they have not affected his chromosomes. So much so, the author goes on to imply, that all the advances on which we so pride ourselves remain biologically precarious, superficial or even exterior to ourselves. There is much that might be said about this; but let us pass over the question of whether we have not undergone some modification, even in our chromosomes, since the era of the pre-Hominids or even that of Cro-Magnon man. Let us concede provisionally that we have developed no hereditary trait in that period rendering us more innately capable of perception and movement in the new dimen-

[6] *Pensées d'un biologiste*, pp. 32–5.

sions of society, space and time. How does this affect our appreciation and evaluation of human progress? I shall show that the answer is splendid and highly encouraging—provided we do not lose sight of the organic reality of the Noosphere.

"Separate the newborn child from human society," you may say, "and you will see how weak he is!" But surely it is clear that this act of isolation is precisely what must not be done, and indeed cannot be done. From the moment when, as I have said, the phyletic strands began to reach toward one another, weaving the first outlines of the Noosphere, a new matrix, coextensive with the whole human group, was formed about the newly born human child—a matrix out of which he cannot be wrenched without incurring mutilation in the most physical core of his biological being. Traditions of every kind, hoarded and manifested in gesture and language, in schools, libraries, museums, bodies of law and religion, philosophy and science—everything that accumulates, arranges itself, recurs and adds to itself, becoming the collective memory of the human race—all this we may see as no more than an outer garment, an epiphenomenon precariously superimposed upon all the other edifices of Nature (the only truly organic ones, as it may appear): but it is precisely this optical illusion which we have to overcome if our realism is to reach to the heart of the matter. It is undoubtedly true that before Man hereditary characteristics were transmitted principally through the reproductive cells. But after the coming of Man another kind of heredity shows itself and becomes predominant; one which was indeed foreshadowed and essayed long before Man, among the highest forms of insects and vertebrates.[7] This is the heredity of example and education. In Man, as though by a

---

[7] A small cynocephalus (baboon), born in captivity, will commit all kinds of blunders when set free (heredity of education). But in similar conditions a young otter, being put in the water, will at once know how to behave (chromosomic heredity). Cf. Eugène N. Marais, *The Soul of the White Ant.*

stroke of genius on the part of Life, and in accord with the grand phenomenon of phyletic coiling, heredity, hitherto primarily chromosomic (that is to say, carried by the genes) becomes primarily "Noospheric"—transmitted, that is to say, by the surrounding environment. In this new form, and having lost nothing of its physical reality (indeed, as much superior to its first state as the Noosphere is superior to the simple, isolated phylum) it acquires, by becoming exterior to the individual, an incomparable substance and capacity. For let me put this question: what system of chromosomes would be as capable as our immense educational system of indefinitely storing and infallibly preserving the huge array of truths and systematized technical knowledge which, steadily accumulating, represents the patrimony of mankind?

Exteriorization, enrichment: we must not lose sight of these two words. We shall come upon them again, quite unchanged, when we turn to consider the machine.

b   The mechanical apparatus. The fact was noted long ago:[8] what has enabled man zoologically to emerge and triumph upon earth, is that he has avoided the anatomical mechanization of his body. In all other animals we find a tendency, irresistible and clearly apparent, for the living creature to convert into tools, its own limbs, its teeth and even its face. We see paws turned into pincers, paws equipped with hooves for running, burrowing paws and muzzles, winged paws, beaks, tusks and so on—innumerable adaptations giving birth to as many phyla, and each ending in a blind alley of specialization. On this dangerous slope leading to organic imprisonment Man alone has pulled up in time. Having arrived at the tetrapod stage he contrived to stay there without further reducing

[8] e.g., Édouard Le Roy, *Les Origines Humaines et le Problème de l'Intelligence.*

the versatility of his limbs. Possessing hands as well as intelligence, and being able, in consequence, to devise artificial instruments and multiply them indefinitely without becoming somatically involved, he has succeeded, while increasing and boundlessly extending his mechanical efficiency, in preserving intact his freedom of choice and power of reason.

The significance and biological function of the tool at last separated from the limb has, as I was saying, long been recognized; and it has long been realized that the tool separated from Man develops a kind of autonomous vitality.[9] We have passive machines giving birth to the active machine, which in turn is followed by the automatic machine. Those are the main classifications; but within each classification what an immense proliferation there is of branches and offshoots, each endowed with a sort of evolutionary potential, irresistible both logically and biologically! We have only to think of the automobile or the airplane.

All this has been noted and often said. But what has not yet been sufficiently taken into account, although it explains everything, is the extent to which this process of mechanization is a collective affair, and the way in which it finally creates, on the periphery of the human race, an organism that is collective in its nature and amplitude.

Let us consider this.

With and since the coming of Man, as we have seen, a new law of Nature has come into force—that of convergence. The convergence of the phyla both ensues from, and of itself leads to, the coming together of individuals within the peculiarly "attaching" atmosphere created by the phenomenon of Reflexion. And out of this

---

[9] e.g., Jacques Lafitte, *Réflexions sur la Science de la Machine. La Nouvelle Journée, no. 21, 1932.*

convergence, as I have said, there arises a very real social inheritance, produced by the synthetic recording of human experience. But if we look for it we may observe precisely the same phenomenon in the case of the machine. Every new tool conceived in the course of history, although it may have been invented in the first place by an individual, has rapidly become the instrument of all men; and not merely by being passed from hand to hand, spreading from one man to his neighbor, but by being adopted corporatively by all men together. What started as an individual creation has been immediately and automatically transformed into a global, quasi-autonomous possession of the entire mass of men. We see this from prehistoric times, and we see it with a vivid clarity in the present era of industrial explosion. Consider the locomotive, the dynamo, the airplane, the cinema, the radio—anything. Can there be any doubt that these innumerable appliances are born and grow, successively and in unison, from roots established in an existing mechanical world-state? For a long time past there have been neither isolated inventors nor machines. To an increasing extent every machine comes into being as a function of every other machine; and, again to an increasing extent, all the machines on earth, taken together, tend to form a single, vast, organized mechanism. Necessarily following the inflexive tendency of the zoological phyla, the mechanical phyla in their turn curve inward in the case of man, thus accelerating and multiplying their own growth and forming a single gigantic network girdling the earth. And the basis, the inventive core of this vast apparatus, what is it if not the thinking center of the Noosphere?

When *Homo faber* came into being the first rudimentary tool was born as an appendage of the human body. Today the tool has been transformed into a mechanized envelope (coherent within itself and immensely varied) appertaining to all mankind. From being somatic it has become "noospheric." And just as the individual at the outset was enabled by the tool to preserve and develop his first, elemental psychic potentialities, so today the Noosphere, disgorging the machine from its innermost organic recesses, is capable of, and in process of, developing a brain of its own.

    c *The cerebral apparatus.* Between the human brain, with its milliards of interconnected nerve cells, and the apparatus of social thought, with its millions of individuals thinking collectively, there is an evident kinship which biologists of the stature of Julian Huxley have not hesitated to examine and expand on critical lines.[10] On the one hand we have a single brain, formed of nervous nuclei, and on the other a Brain of brains. It is true that between these two organic complexes a major difference exists. Whereas in the case of the individual brain thought emerges from a system of nonthinking nervous fibers, in the case of the collective brain each separate unit is in itself an autonomous center of reflection. If the comparison is to be a just one we must, at every point of resemblance, take this difference into account. But when all allowance is made the fact remains that the analogies between the two systems are so numerous, and so compelling, that reason forbids us to regard the parallel as either purely superficial or a mere matter of chance. Let us take a rapid glance at the structure and functioning of what might be termed the "cerebroid" organ of the Noosphere.

    First the structure: and here I must turn back to the machine. I have said that, thanks to the machine, Man has contrived both severally and collectively to prevent the best of himself from being absorbed in purely physiological and functional uses, as has happened to other animals. But in addition to its protective note, how can we fail to see the machine as playing a constructive part in the creation of a truly collective consciousness? It is not merely a matter of the machine which liberates, relieving both individual and

[10] Lecture delivered in New York and published in the *Scientific Monthly*, 1940.

collective thought of the trammels which hinder its progress, but also of the machine which creates, helping to assemble, and to concentrate in the form of an ever more deeply penetrating organism, all the reflective elements upon earth.

I am thinking, of course, in the first place of the extraordinary network of radio and television communications which, perhaps anticipating the direct syntonization of brains through the mysterious power of telepathy, already link us all in a sort of "etherized" universal consciousness.

But I am also thinking of the insidious growth of those astonishing electronic computers which, pulsating with signals at the rate of hundreds of thousands a second, not only relieve our brains of tedious and exhausting work but, because they enhance the essential (and too little noted) factor of "speed of thought," are also paving the way for a revolution in the sphere of research. And there are other forms of technical equipment, such as the electronic microscope whereby our sensory vision, the principal source of our ideas, has been enabled to leap the optical gap between the cell and the direct observation of large molecules.

There is a school of philosophy which smiles disdainfully at these and kindred forms of progress. "Commercial machines," we hear them say, "machines for people in a hurry, designed to gain time and money." One is tempted to call them blind, since they fail to perceive that all these material instruments, ineluctably linked in their birth and development, are finally nothing less than the manifestation of a particular kind of super-Brain, capable of attaining mastery over some supersphere in the universe and in the realm of thought. "Everything for the individual!"—such was the reaffirmation of my brilliant friend, Gaylord Simpson,[11] in a recent outburst

[11] George Gaylord Simpson, "The Role of the Individual in Evolution," *Journal of the Washington Academy of Sciences*, vol. 31, no. 1, 1941.

of antitotalitarian fervor. But let us grasp this point clearly. No doubt it is true, scientifically speaking, that no distinct center of superhuman consciousness has yet appeared on earth (at least in the *living world*) for which it may be claimed or predicted that one day it will exercise a centralizing function, in relation to associated human thought, similar to the role of the individual "I" in relation to the cells of the brain. But that is far from saying that, influenced by the links which unite them, our grouped minds working together are not capable of achieving results which no one member of the group could achieve alone, and from which every individual within the collective process benefits "integrally," although still not in the total sense.

We have only to consider any of the new concepts and intuitions which, particularly during the past century, have become or are in process of becoming the indestructible keystones and fabric of our thought—the idea of the atom, for example, or of organic Time or Evolution. It is surely obvious that no man on earth could alone have evolved them; no one man, *thinking by himself*, can encompass, master or exhaust them; yet every man on earth shares, *in himself*, in the universal heightening of consciousness promoted by the existence in our minds of these new concepts of matter and new dimensions of cosmic reality. It is not a question of simple repetitive "summation" but of synthesis. Not, it is true (at least not yet, here below) synthesis pushed to the point where it calls into being some new kind of autonomous supercenter in the depths of the synthesized, but a synthesis which at least suffices to erect, as though it were a vault above our heads, a sphere of mutually reinforced consciousness, the seat, support and instrument of supervision and superideas. No doubt everything proceeds from the individual and in the first instance depends on the individual; but it is on a higher level than the individual that everything achieves its fulfillment.

We have touched upon the apparatus of heredity, machinery and mind. It would be rash and often absurd to attempt to pursue further, and in detail, the comparison between the organism of the individual and that of the Noosphere. But the fact that the general line of analogy is valid and fruitful seems to me to be definitely proved by the very remarkable fact that these three systems, taken in conjunction, not only form a complementary and coherent whole, consistent within itself, but, which is even more easy of demonstration, that this whole is capable of breaking into motion and of working—that it *functions*, in a word.

3. *The Physiology of the Noosphere*

ONE OF THE most impressive effects of the power of collective vision which is conferred upon us by the formation of a common brain is the perception of "great slow movements," so vast and slow that they are only observable over immense stretches of time. The currents that give birth to sidereal systems; the folds and upthrusts that form mountains and continents; the ebb and flow within the biosphere—in each case what we had supposed to be the extreme of immobility and stability is discovered to be a state of fundamental and irresistible movement.

So it is with the Noosphere.

I have already attempted a sort of anatomy of the major organs of the Noosphere. It remains for me to show that these separate parts, planetary in their dimensions, are not designed to remain in a state of rest. The formidable wheels turn, and in their combined action hidden forces are engendered which circulate throughout the gigantic system. What goes on around us in the human mass is not merely a flurry of disordered movement, as in a gas; something is purposefully stirring, as in a living being.

Let us try to gain some understanding of this vast internal process of which we are all a part and to which we all contribute, almost without knowing it.

At the heart of the entire movement, like the mainspring of a clock, there reappears, in identifiable form, what we have termed the *inflexion* of human stems upon themselves. It was of this mysteriously compelled in-folding, as I have said, that the human race was born. I will now add that it is through the continued operation of the same movement that the race persists and functions. Indeed, we have only to open our eyes to be as it were spellbound by the dazzling vision, the spectacle of human shoots caught in the combined play of irresistible forces which slowly but surely continue to close and coil about us. Despite the havoc of war, the population on the limited surface of this planet which bears us is increasing in almost geometrical progression; while at the same time the scope of each human molecule, in terms of movement, information and influence, is becoming rapidly coextensive with the whole surface of the globe. A state of tightening compression, in short; but even more, thanks to the biological intermingling developed to its uttermost extent by the appearance of Reflection, a state of organized compenetration, in which each element is linked with every other. No one can deny that a network (a world network) of economic and psychic affiliations is being woven at ever-increasing speed which envelops and constantly penetrates more deeply within each of us. With every day that passes it becomes a little more impossible for us to act or think otherwise than collectively.

What is the significance of this multiform embrace, both external and internal, against which we struggle in vain? Can it mean that, caught in the ramifications of a sightless mechanism, we are destined to perish by stifling each other? No; for as the coil grows tighter and the tension rises the forces of supercompression in the vast generator find an effective outlet.

We begin to catch sight of it in the study of an all too familiar phenomenon, disquieting in appearance, but in fact highly revealing and reassuring—the *phenomenon of unemployment.* Owing to the extraordinarily rapid development of the machine a rapidly increasing number of workers, running into tens of millions, are out of work. The experts gaze in dismay at this economic apparatus, their own creation, which instead of absorbing all the units of human energy with which they furnish it rejects an increasing number, as though the machine they devised were working to defeat their purpose. Economists are horrified by the growing number of idle hands. Why do they not look a little more to biology for guidance and enlightenment? In its progress through a million centuries, mounting from the depths of the unconscious to consciousness, when has Life proceeded otherwise than by releasing psychic forces through the medium of the mechanisms it has devised? We have only to consider the evolution of the nervous system in the animal series, proceeding by chronological stages over a great period of time. Or let the theorists consider themselves. How are they capable of reasoning at all if not because within them their visceral system has been taught to function automatically, while around them society is so well organized that they have both the strength and the leisure to calculate and reflect? What is true for each individual man is precisely what is happening at this moment on the higher level of mankind. Like a heavenly body that heats as it contracts, such, and in a twofold respect, is the Noosphere: first in intensity, the degree in which its tension and psychic temperature are heightened by the coming together and mutual stimulation of thinking centers throughout its extent; and also quantitatively through the growing number of people able to use their brains because they are freed from the need to labor with their hands. So that to attempt to suppress unemployment by incorporating the unemployed in the machine would be against the purpose of Nature and a biological absurdity. The Noosphere can function only by releasing more and more spiritual energy with an ever higher potential.

To all this you may remark as follows: "Very well; let us agree that the combined effect of phyletic intertwining and mechanical progress causes life to boil over. But in that case, and surely it is the root of the matter, how are we to canalize and use the rising tide of liberated consciousness, that is still so crude and unformed?" My answer is: "By transforming it." And at this point, having invited you to reflect upon the phenomenon of unemployment, I will draw your attention to another and no less universal phenomenon, equally characteristic of the present age—*the phenomenon of research.*

Understanding, discovery, invention . . . From the first awakening of his reflective consciousness, Man has been possessed by the demon of discovery; but until a very recent epoch this profound need remained latent, diffused and unorganized in the human mass. In every past generation true seekers, those by vocation or profession, are to be found; but in the past they were no more than a handful of individuals, generally isolated, and of a type that was virtually abnormal—the "inquisitive." Today, without our having noticed it, the situation is entirely changed. In fields embracing every aspect of physical matter, life and thought, the research-workers are to be numbered in hundreds of thousands, and they no longer work in isolation but in teams endowed with penetrative powers that it seems nothing can withstand. In this respect too, the movement is becoming generalized and is accelerating to the point where we must be blind not to see in it an essential trend in human affairs. Research, which until yesterday was a luxury pursuit, is in process of becoming a major, indeed the principal, function of humanity. As to the significance of this great event,

I for my part can see only one way to account for it. It is that the enormous surplus of free energy released by the in-folding of the Noosphere is destined by a natural evolutionary process to flow into the construction and functioning of what I have called its "Brain." As in the case of all the organisms preceding it, but on an immense scale, humanity is in process of "cerebralizing" itself. And our proper biological course, in making use of what we call our leisure, is to devote it to a new kind of work on a higher plane: that is to say, to a general and concerted effort of vision. The Noosphere, in short, is a stupendous thinking machine.

It is in this sense alone, as I believe, that the horizon appears and we can gain a clear view of the human world surrounding us. In harmony with the cosmic impulse which leads to the constant disintegration of atoms and the attendant release of energy, Life (though probably localized on a few rare planets) compels us increasingly to view it as an underlying current in the flow of which matter tends to order itself upon itself with the emergence of consciousness. On the one hand we have physical radiation bound up with disintegration; and on the other hand psychic radiation bound up with an ordered aggregation of the stuff of the universe. In the eyes of nineteenth-century science the interiorization of the world, leading to the phenomenon of Reflection, might still pass for an accident and an anomaly. We now see it to be a clearly defined process coextensive with the whole of reality. Complexification due to the growth of consciousness, or consciousness the outcome of complexity: experimentally the two terms are inseparable. Like a pair of related quantities they vary simultaneously. And surely it is within this generalized cosmic process that the Noosphere, a particular and extreme case, has its natural place and takes its shape. The maximum of complication, represented by phyletic in-folding, and in consequence the maximum of consciousness emerging from the system of individual brains, coordinated and mutually supporting. And this is exactly what was to be expected.

But it is assuredly a remarkable coincidence that in justifying the organic interpretation of the Phenomenon of Man, as we have sought to do, we should also be paving the way for a reasonable forecast as to our future destiny, and the fate which is reserved for us at the end of Time.

4. The Phases and Future of the Noosphere

WE HAVE FOUND it possible to express the social totalization which we are undergoing in terms of a clearly identifiable biological process: proceeding from this we may surely look into the future and predict the course of the trajectory we are describing. Once we have accepted that the formation of a collective human organism, a Noosphere, conforms to the general law of recurrence which leads to the heightening of Consciousness in the universe as a function of complexity, a vast prospect opens before us. To what regions and through what phases may we suppose that the extension of the rising curve of hominization will carry us?

Immediately confronting us (indeed, already in progress) we have what may be called a "phase of planetization."

It can truly be said, no doubt, that the human group succeeded long ago in covering the face of the earth, and that over a long period its state of zoological ubiquity has tended to be transformed into an organized aggregate; but it must be clear that the transformation is only now reaching its point of full maturity. Let us glance over the main stages of this long history of aggregation. First, in the depths of the past, we find a thin scattering of hunting groups spread here and there throughout the Ancient World. At a later stage, some fifteen thousand years ago, we see a second scattering, very much more dense and clearly defined: that of agricultural groups installed in fertile valleys—centers of social life where man, arrived at a state of stability, achieved the expansive powers which were to enable him to invade the New World. Then, only seven or eight thousand years ago, there came the first civilizations, each covering a large part of a continent. These were succeeded by the real empires. And so on . . . patches of humanity growing steadily larger, overlapping, often absorbing one another, thereafter to break apart and again reform in still larger patches. As we view this process, the spreading, thickening and irresistible coalescence, can we fail to perceive its eventual outcome? The last blank spaces have vanished from the map of mankind. There is contact everywhere, and how close it has become! Today, embedded in the economic and psychic network which I have described, two great human blocks alone remain confronting one another. Is it not inevitable that in one way or another these two will eventually coalesce? Preceded by a tremor, a wave of "shared impulse" extending to the very depths of the social and ethnic masses in their need and claim to participate, without distinction of class or color, in the onward march of human affairs, the final act is already visibly preparing. Although the form is not yet discernible, mankind tomorrow will awaken to a "panorganized" world.

But, and we must make no mistake about this, there will be an essential difference, a difference of order, between the unitary state toward which we are moving and everything we have hitherto known. The greatest empires in history have never covered more than fragments of the earth. What will be the specifically new manifestations which we have to look for in the transition to *totality*? Until now we have never seen mind manifest itself on this planet except in separated groups and in the static state. What sort of current will be generated, what unknown territory will be opened up, when the circuit is suddenly completed?

I believe that what is now being shaped in the bosom of planetized humanity is essentially a *rebounding* of evolution upon itself. We all know about the real or imaginary projectiles whose impetus is renewed by the firing of a series of staged rockets. Some such procedure, it seems to me, is what Life is preparing at this moment, to accomplish the supreme, ultimate leap. The first stage was the elaboration of lower organisms, up to and including Man, by the use and irrational combination of elementary sources of energy received or released by the planet. The second stage is the superevolution of Man, individually and collectively, by the use of refined forms of energy scientifically harnessed and applied in the bosom of the Noosphere, thanks to the coordinated efforts of all men working reflectively and unanimously upon themselves. Who can say whither, coiled back upon our own organism, our combined knowledge of the atom, of hormones, of the cell and the laws of heredity will take us? Who can say what forces may be released, what radiations, what new arrangements never hitherto attempted by Nature, what formidable powers we may henceforth be able to use, for the first time in the history of the world? This is Life setting out upon a second adventure from the springboard it established when it created humankind.

But all this is no more than the outward face of the phenomenon. In becoming planetized humanity is acquiring new physical powers which will enable it to superorganize matter. And, even more important, is it not possible that by the direct converging of its members it will be able, as though by resonance, to release psychic powers whose existence is still unsuspected? I have already spoken of the recent emergence of certain new faculties in our minds, the sense of genetic duration and the sense of collectivity. Inevitably, as a natural consequence, this awakening must enhance in us, from all sides, a generalized sense of the organic, through which the entire complex of interhuman and intercosmic relations will become charged with an immediacy, an intimacy and a realism such as has long been dreamed of and apprehended by certain spirits particularly endowed with the "sense of the universal," but which has never yet been *collectively applied*. And it is in the depths and by grace of this new inward sphere, the attribute of planetized Life, that an event seems possible which has hitherto been incapable of realization: I mean the pervasion of the human mass by the power of sympathy. It may in part be passive sympathy, a communication of mind and spirit that will make the phenomenon of telepathy, still sporadic and haphazard, both general and normal. But above all it will be a state of active sympathy in which each separate human element, breaking out of its insulated state under the impulse of the high tensions generated in the Noosphere, will emerge into a field of prodigious affinities, which we may already conjecture in theory. For if the power of attraction between simple atoms is so great, what may we not expect if similar bonds are contracted between human molecules? Humanity, as I have said, is building its composite brain beneath our eyes. May it not be that tomorrow, through the logical and biological deepening of the movement drawing it together, it will find its *heart*, without which the ultimate wholeness of its powers of unification can never be fully achieved? To put it in other words, must not the constructive developments now taking place within the Noosphere in the realm of sight and reason necessarily also penetrate to the sphere of feeling? The idea may seem fantastic when one looks at our present world, still dominated by the forces of hatred and repulsion. But is not this simply because we refuse to heed the admonitions of science, which is daily proving to us, in every field, that seemingly impossible changes become easy and even inevitable directly there is a change in the order of the dimensions?

To me two things, at least, now seem certain. The first is that, following the state of collective organization we have already achieved, the process of planetization can only advance ever further in the direction of growing unanimity. And the second is that this growth of unanimity, being of its nature convergent, cannot continue indefinitely without reaching the natural limit of its course. Every cone has an apex. In the case of this human aggregation how shall we seek, not to imagine but to define the supreme point of coalescence? In terms of the strictly phenomenal viewpoint which I have adopted throughout this paper, it seems to me that the following may be said:

What at the very beginning made the first man, was, as we know, the heightening of the individual consciousness to the point where it acquired the power of Reflection. And the measure of human progress during the centuries which followed is, as I have sought to show, the increase of this reflective power through the interaction, or conjugated thought, of conscious minds working upon one another. Well, what will finally crown and limit collective humanity at the ultimate stage of its evolution, is and must be, by reason of continuity and homogeneity, the establishment of a sort of focal point at the heart of the reflective apparatus as a whole.

If we concede this the whole of human history appears as a progress between two critical points: from the lowest point of elementary consciousness to the ultimate, noospherical point of Reflection. In biological terms, humanity will have completed itself and fully achieved its internal equilibrium only when it is psychically centered upon itself (which may yet take several million years).

In a final effort of thought let us remove ourselves to that ultimate summit where in the remote future, *but seen from the present*, the tide which bears us reaches its culmination. Is there anything further to be discerned beyond that last peak etched against the horizon?—Yes and no.

In the first place no, because at that mysterious pole crowning our ascent the compass that has guided us runs amok. It was by the law of "consciousness and complexity" that we set our course: a consciousness becoming ever more centered, emerging from the heart of an increasingly vast system of more numerous and better organized elements. But now we are faced by an entirely new situation: for the first time we have no multiple material under our hands. Unless, as seems infinitely improbable, we are destined by contact with other thinking planets, across the abysses of space and time, some day to become integrated within an organized complex composed of a number of Noospheres, humanity, having reached maturity, will remain alone, face to face with itself. And at the same time our law of recurrence, based on the play of interrelated syntheses, will have ceased to operate.

So in one sense it all seems to be over; as though, having reached its final point of Noospheric Reflexion, the cosmic impulse toward consciousness has become exhausted, condemned to sink back into the state of disintegration implacably imposed on it by the laws of stellar physics. But in another sense nothing will be ended: for at this point, and at the height of its powers, individual consciousness acquires the formidable property something else comes into operation, a primary attribute of Reflection concerning which we have hitherto said nothing—*the will to survive*. In reflecting upon itself the individual consciousness acquires the formidable property of foreseeing the future, that is to say, death. And at the same time it knows that it is psychologically impossible for it to continue to work in pursuance of the purposes of Life unless something, the best of the work, is preserved from total destruction. In this resides the whole problem of action. We have not yet taken sufficient account of the fact that this demand for the Absolute, not always easily discernible in the isolated human unit, is one of the impulses which grow and are intensified in the Noosphere. Applied to the individual the idea of total extinction may not at first sight appall us; but extended to humanity as a whole it revolts and sickens us. The fact is that the more Humanity becomes aware of its duration, its number and its potentialities—and also of the enormous burden it must bear in order to survive—the more does it realize that if all this labor is to end in nothing, then we have been cheated and can only rebel. In a planetized Humanity the *insistence upon irreversibility* becomes a specific requisite of action; and it can only grow and continue to grow as Life reveals itself as being ever more rich, an ever heavier load. So that, paradoxically, it is at that ultimate point of centration which renders it cosmically unique, that is to say apparently incapable of any further synthesis, that the Noosphere will have become charged to the fullest extent with psychic energies to impel it forward in yet another advance. . . .

And what can this mean except that, like those planetary orbits which seem to traverse our solar system without remaining within it, the curve of consciousness, pursuing its course of growing complexity, will break through the material framework of Time and Space to escape somewhere toward an ultracenter of unification and consistence, where there will finally be assembled, and in detail, everything that is irreplaceable and incommunicable in the world.

And it is here, an inevitable intrusion in terms of biology, and in its proper place in terms of science, that we come to the problem of God.

Conclusion: The Rise of Freedom

LET US TURN to cast an eye over the road that we have followed.

At the beginning we seemed to see around us nothing but a disconnected and disordered humanity: the crowd, the mass, in which, it may be, we saw only brutality and ugliness. I have tried, fortified by the most generally accepted and solid conclusions of science, to take the reader above this scene of turmoil; and as we have risen higher so has the prospect acquired a more ordered shape. Like the petals of a gigantic lotus at the end of the day, we have seen human petals of planetary dimensions slowly closing in upon themselves. And at the heart of this huge calyx, beneath the pressure of its in-folding, a center of power has been revealed where spiritual energy, gradually released by a vast totalitarian mechanism, then concentrated by heredity within a sort of super-brain, has little by little been transformed into a common vision growing ever more intense. In this spectacle of tranquillity and intensity, where the anomalies of detail, so disconcerting on our individual scale, vanish to give place to a vast, serene and irresistible movement from the heart, everything is contained and everything harmonized in accord with the rest of the universe. Life and consciousness are no longer chance anomalies in Nature; rather we find in biology a complement to the physics of matter. On the one hand, I repeat, the stuff of the world dispersing through the radiation of its elemental energy; and on the other hand the same stuff reconverging through the radiation of thought. The fantastic at either end: but surely the one is necessary to balance the other? Thus harmony is achieved in the ultimate perspective, and, furthermore, a program for the future: for if this view is accepted we see a splendid goal before us, and a clear line of progress. Coherence and fecundity, the two criteria of truth.

Is this all illusion, or is it reality?

It is for the reader to decide. But to those who hesitate, or who refuse to commit themselves, I would say: "Have you anything else, anything better to suggest that will account scientifically for the phenomenon of man considered as a whole, in the light of his past development and present progress?"

You may reply to me that this is all very well, but is there not something lacking, an essential element, in this system which I claim to be so coherent? Within that grandiose machine-in-motion which I visualize, what becomes of that pearl beyond price, our personal being? What remains of our freedom of choice and action?

But do you not see that from the standpoint I have adopted it appears everywhere—and is everywhere heightened?

I know very well that by a kind of innate obsession we cannot rid ourselves of the idea that we become most masters of ourselves by being as isolated as possible. But is not this the reverse of the truth? We must not forget that in each of us, by our very nature, everything is in an elemental state, including our freedom of action. We can only achieve a wider degree of freedom by joining and associating with others in an appropriate way. This is, to be sure, a dangerous operation, since, whether it be a case of disorderly intermingling, or of some simple form of coordination, like the meshing of gear-wheels, our activities tend to cancel one another out or to become mechanical—we find this only too often in practice. Yet it is also salutary, since the approach of spirit to spirit in a common vision or a shared passion undoubtedly enriches all; in the case of a team, for example, or of two lovers. Achieved with sympathy, union does not restrict but exalts the possibilities of our being. We see this everywhere and every day on a limited scale. Why should it not be worth correspondingly more on a vast and all-embracing scale, if the law applies to the very structure of things? It is simply a question of tension within the field that polarizes and attracts. In the case of a blind aggregation, of some form of purely mechanical arrangement, the effect of large numbers is to materialize our activities. That is true: but where it is a matter of unanimity realized from *within* the effect is to personalize them, and, I will add, to make them unerring. A single freedom, taken in isolation, is weak and uncertain and may easily lose itself in mere groping. But a totality of freedom, freely operating, will always end by finding its road. And this incidentally is why throughout this paper, without seeking to minimize the uncertainties inherent in Man's freedom of choice in relation to the world, I have been able implicitly to maintain that we are moving both freely and ineluctably in the direction of concentration by way of planetization. One might put it that determinism appears at either end of the process of cosmic evolution, but in antithetically opposed forms: at the lower end it is forced along the line of the most probable *for lack of freedom*; at the upper end it is an ascent into the improbable through *the triumph of freedom*.

We may be reassured. The vast industrial and social system by which we are enveloped does not threaten to crush us, neither does it seek to rob us of our soul. The energy emanating from it is free not only in the sense that it represents forces that can be used: it is moreover free because, in the whole no less than in the least of its elements, it arises in a state that is ever more spiritualized. A thinker such as Cournot[12] might still be able to suppose that the socialized group degrades itself biologically in terms of the individuals which comprise it. Only by reaching to the heart of the Noosphere (we see it more clearly today) can we hope, and indeed be sure, of finding, all of us together and each of us separately, the fullness of our humanity.

REVUE DES QUESTIONS SCIENTIFIQUES (LOUVAIN),

JANUARY 1947, PP. 7–35.

---

[12] Cournot, *Considérations sur la Marche des idées et des Événements dans les Temps modernes.* (Réédition Mentré. Vol. II, p. 178).

CHAPTER 11

FAITH IN MAN

1. Definition and Novelty

BY "FAITH IN Man" we mean here the more or less active and fervent conviction that Mankind as an organic and organized whole possesses a future: a future consisting not merely of successive years but of higher states to be achieved by struggle. Not merely survival, let us be clear, but some form of higher life or superlife.

Considered in its deepest origins this human trend toward a state of higher being is as old and universal as the world itself. As far back as we can trace it, and even in its humblest manifestations, the advance of Life, however spurred on by the sheer, hard necessity of continuing to exist, has always been inspired by an expectation of something greater. Are not Nature's countless experiments all variants of a single act of faith, an obstinate feeling of the way toward an outlet leading forward and ever higher? Above all, at that critical point where instinct turned reflexively to thought, and awareness of the future became an accomplished fact on earth, must not Man, in whom this radical change occurred, even in his most primitive state have experienced the vital urge to grasp all things and transcend himself?

Mythology and folklore (we shall come back to this) are, in fact, filled with symbols and fables expressing the deeply rooted resolve of Earth to find its way to Heaven; from which it follows that we may in a perfectly legitimate sense accept the fact that a generalized, implicit faith of Man in Man is older than all civilization, and that it is this, finally, which constitutes the basic impulse informing all our past history.

But is there not another and even truer sense in which we must affirm that this faith, in the explicit, collective form of our definition, represents a specific new attitude in the world and therefore calls for our particular attention?

I believe that this is so, on the following grounds.

A major problem posed by the fact, of which we are henceforth assured, that the Universe is in a state of psychic evolution, is the question of how far its evolutionary course is likely to affect our future power of thought. Whatever the eventual answer may be, two things are undeniable: first, that at certain moments in the past, human consciousness—however unchanging in its essential framework—has risen to the perception of new dimensions and values; and secondly that the age in which we are living is precisely such a moment of awakening and transformation. In the course of a few generations, almost without our realizing it, our view of the world has been profoundly altered. Under the combined influence of Science and History, and of social developments, the twofold sense of duration and collectivity has pervaded and reordered the entire field of our experience; with the twofold result that the future, hitherto a vague succession of monotonous years awaiting an unimportant number of scattered individual lives, is now seen to be a period of positive becoming and maturing—but one in which we can advance and shape ourselves only in solidarity.

Thus we have the simultaneous growth in our minds of two essentially modern concepts, those of collectivity and of an organic
future: a double development precisely engendering the deep-
rooted change of heart that was required to bring about the direct
transformation of a childlike and instinctive faith in Man into its
rational, adult state of constructive, militant faith in Mankind!

A spiritual crisis was inevitable: it has not been slow in coming.

But let us look with open minds at the new world being born
around us amid the convulsions of war. Disregarding the superficial
chaos which prevents us from seeing clearly, probing beneath the un-
speakable disorders that so dismay us, let us try to take the pulse and
temperature of Earth. If we have any power to diagnose we are
bound to recognize that the so-called ills which so afflict us are above
all growing pains. What looks like no more than a hunger for material well-being is in reality a hunger for higher being: it is the spirit of
Mankind suddenly alive with the sense of all that remains to be done
if it is to achieve the fulfillment of its powers and possibilities.

2. *Power and Ambiguity*

IT WOULD BE criminal or insane to attempt to resist the great explosion of the inner forces of the Earth that is now beginning. Like
the collectivization which accompanies it, this upsurge of human
faith which we are witnessing is a life-bearing phenomenon, and
therefore irresistible. But that does not mean that we should let ourselves be borne passively and indiscriminately on the tide. The
more youthful and forceful the energy, the more misguided and
dangerous may be its ebullience. We see this all too clearly in the
present-day world.

We sincerely believe that in itself, and in its only legitimate and
enduring form, faith in Man does not exclude but must on the con-

trary include the worship of Another—One who is higher than Man. To grow in stature and strength so as to be able to give more of oneself and clasp in a tighter embrace (as in the Bible story of Jacob wrestling with the Angel; and as happens on an everyday level in every passionate union), this is the true and noble manner of interpreting and canalizing the impulse which urges us upward.

But, as the facts prove only too well, this first way of believing in Man goes hand in hand with another way, more elementary, immediate and simple, and therefore more alluring. Correctly interpreted, I repeat, faith in Man can and indeed must cast us at the feet and into the arms of One who is greater than ourselves. But, it can be argued, why after all should we not conceive this One who is greater than ourselves as being in fact identical with ourselves? Given the power he possesses, why should Man look for a God outside himself? Man, self-sufficient[1] and wholly autonomous, sole master and disposer of his destiny and the world's—is not this an even nobler concept?

Here we have the modern version of the heroic temptation of all time, that of the Titans, of Prometheus, of Babel and of Faust; that of Christ on the mountain; a temptation as old as Earth itself; as old as the first reflective awakening of Life to the awareness of its powers. But it is a temptation which is only now entering its critical phase, now that Man has raised himself to the point of being able to measure both the immensity of the Time that lies before him and the almost limitless powers made available to him by his concerted efforts to seize hold of the material springs of the world.

Is the dilemma insoluble or (as we would rather believe) only a temporary one, destined to vanish like so many others when we have reached a higher level of spiritual evolution? We may be in two minds about this.

---

[1] Teilhard uses the English word.

The fact remains that at the present time a fundamental inner impulse, newly born in our hearts, is tending to find a dual, and divergent, expression in two apparently incompatible spiritual forms; on the one hand, the spirit (let us call it "Christian") of sacrifice and of union centered in the expectation of a Vision in the future; and on the other hand the Promethean or Faustian spirit of self-worship based on the material organization of the earth. The ambiguity is there. And because (always by virtue of a rhythm which may be reversed tomorrow) it is the material and tangible aspect which at this moment of world history seems to hold the initiative in the advance of Life, the struggle is proceeding in a way which suggests that the Promethean faith is the only valid one, or at least the more active. We see no other in the service of the world, or we run the risk of seeing no other. Hence the tendency (which is also as old as the world) of the defenders of the Spirit to regard as diabolical, and to reject as being among the most formidable manifestations of pride, the irrepressible desire for growth and conquest, the unshakeable sense of power and progress, which at present fills the human breast.

But we must not leap to conclusions. Since by definition ambiguity is not perversity but only the danger of perversion, which after all is not the same thing, let us seek to place ourselves psychologically at a level below the point where the dilemma seems to be resolving itself in two irreconcilable forms. In other words, let us try to understand what faith in Man signifies in its undifferentiated state (pre-Promethean or pre-Christian); what it looks for and what it offers us.

3. *The Uniting Force*

PRESENT-DAY MANKIND, as it becomes increasingly aware of its unity—not only past unity in the blood, but future unity in progress—is experiencing a vital need to close in upon itself. A tendency toward unification is everywhere manifest, and especially in the different branches of religion. We are looking for something that will draw us together, below or above the level of that which divides. It may be said, in the aftermath of the war, that this need is spontaneously and unanimously arising on every hand. But where are we to discover the mysterious principle of *rapprochement?* Are we to look downward or upward—to our common interest or our common faith?

We must by no means underestimate the force of common interest in a matter of this sort. The visible success of communal undertakings in which the material well-being of the individual becomes essentially dependent on the functioning of the association as a whole; more still, on the world scale, the example of the last war, in which a common danger for a time welded together large sections of the world—all this decidedly proves that physical necessity, when it happens to coincide, is a synthesizing factor between human particles. But this kind of synthesis, we must note, remains fragile in two respects: firstly, because the coincidence which brought it about is in the nature of things temporary and accidental; secondly, and above all, because elements brought together under the compulsion of necessity or fear cohere only outwardly and on the surface. When the wave of fear or common interest has passed, the union dissolves without having given birth to a soul. Not through external pressure but only from an inward impulse can the unity of Mankind endure and grow.

And this, it seems, is where the major, "providential" role reserved by the future for what we have called "faith in Man" displays itself. A profound common aspiration arising out of the very shape of the modern world—is not this specifically what is most to be desired, what we most need to offset the growing forces of dissolution and dispersal at work among us?

But here we must be on our guard.

Recently, and in particular through the sympathetic pen of Aldous Huxley, an effort has been made to formulate and crystallize, in a series of abstract propositions, the basis of a common philosophy on which all men of goodwill can agree in order that the world may continue to progress. We believe this to be helpful, and moreover we are persuaded that gradually, in religious thought as in the sciences, a core of universal truth will form and slowly grow, to be accepted by everyone. Can there be any true spiritual evolution without it? But shall we not be misled by this formulation of a common view of the world, infinitely precious in itself, if we consider it simply in terms of its application and result, without looking for the principle and fecundating act of a genuine union? Any abstract scheme tends of its nature to resolve in an arbitrary fashion, and perhaps prematurely for the whole, the ambiguity of the future. There is the risk that it will restrict the movement to a given direction, whereas it is out of the movement as such that the desired effect of unification must come.

But at the youthful stage in which we are at present considering it, Faith in Man proceeds and operates in a quite different fashion.

It is true that at the outset it presupposes a certain fundamental concept of the place of Man in Nature. But as it rises above this rationalized common platform it becomes charged with a thousand differing potentialities, elastic and even fluid—indivisible, one might say, by the expressions of hostility to which Thought, in its gropings, may temporarily subject it. Indivisible and even triumphant: for despite all seeming divisions (this is what matters) it continues unassailably to draw together and even to reconcile everything that it pervades. Take the two extremes confronting us at this moment, the Marxist and the Christian, each a convinced believer in his own particular doctrine, but each, we must suppose,

fundamentally inspired with an equal faith in Man. Is it not in-
contestable, a matter of everyday experience, that each of these, to
the extent that he believes (and sees the other believe) in the future
of the world, feels a basic human sympathy for the other—not for
any sentimental reason, but arising out of the obscure recognition
that both are going the same way, and that despite all ideological
differences they will eventually, in some manner, come together on
the same summit? No doubt each in his own fashion, following his
separate path, believes that he has once and for all solved the riddle of the world's future. But the divergence between them is in reality neither complete nor final, unless we suppose that by some
inconceivable and even contradictory feat of exclusion (contradictory because nothing would remain of his faith) the Marxist, for
example, were to eliminate from his materialistic doctrine every
upward surge toward the spirit. Followed to their conclusion the
two paths must certainly end by coming together: for in the nature
of things everything that is faith must rise, and everything that rises
must converge.

In short we may say that faith in Man, by the combined effect
of its universality and its elemental quality, shows itself upon examination to be the general atmosphere in which the higher, more
elaborated forms of faith which we all hold in one way or another
may best (indeed *can only*) grow and come together. It is not a *formula*, it is the *environment* of union.

No one can doubt that we are all more or less affected by this
elementary, primordial faith. Should we otherwise truly belong to
our time? And if, through the very force of our spiritual aspirations, we have been inclined to mistrust it, even to feel that we are
immune from it, we must look more closely into our own hearts. I
have said that the soul has only one summit. But it has also only
one foundation. Let us look well and we shall find that our Faith in

God, detached as it may be, sublimates in us a rising tide of human aspirations. It is to this original sap that we must return if we wish to communicate with the brothers with whom we seek to be united.

ADDRESS TO THE WORLD CONGRESS OF FAITHS
(FRENCH SECTION), MARCH 8, 1947.

CHAPTER 12

SOME REFLECTIONS ON
THE RIGHTS OF MAN

AS FIRST PROCLAIMED, in 1789, the Rights of
Man were primarily an expression of the individual will to autonomy—"Everything for the Individual within Society"—implying that the human race
was designed to unfold and culminate in a multiplicity of units achieving, each in itself, their maximum development. This seems to have been the
ruling preoccupation and vision of the eighteenth-
century humanitarians.

Since then, however, owing to the increasing
importance of the various forms of collectivity in
human society, the nature of the problem has profoundly changed. We can no longer doubt this. For
innumerable convergent reasons (the rapid increase of ethnic, economic, political and cultural
links) the human individual finds himself defini-
tively involved in an irresistible process tending
toward a system of organopsychic solidarity on
earth. Whether we wish it or not, Mankind is becoming collectivized, totalized under the influence
of psychic and spiritual forces on a planetary scale.
Out of this has arisen, in the heart of every man,

the present-day conflict between the individual, ever more conscious of his individual worth, and social affiliations which become ever more demanding.

But the conflict, if we think of it, is only one of appearance. Biologically, as we know, the human unit is not self-sufficing. In other words it is not in isolation (as we might have supposed) but only in *appropriate* association with his fellows that the individual can hope to attain to the fullness of his *personality*, his energies, his power of action and his consciousness, more especially since we do not become completely "reflective" (that is to say, "men") except by being reflected in each other. Collectivization and individualization (in the sense of personality, not of social autonomy) are thus not opposed principles. The problem is so to order matters as to ensure that human totalization is brought about, not by the pressure of external forces, but through the internal workings of harmonization and sympathy.

It at once becomes clear, when we adopt this altered standpoint, that the purpose of a new Declaration of the Rights of Man cannot be, as formerly, to ensure the highest possible degree of independence for the individual in society, but to define the conditions under which the inevitable totalization of Mankind may be effected, not only without impairing but so as to enhance, I will not say the autonomy of each of us but (a quite different thing) the incommunicable singularity of being which each of us possesses.

We must no longer seek to organize the world in favor of, and in terms of, the isolated individual; we must try to combine all things for the perfection ("personalization") of the individual by his well-ordered integration with the unified group in which Mankind must eventually culminate, both organically and spiritually. That is the problem.

Thus transposed into the framework of an operation with two variables (the progressive, interdependent adjustment of the two processes of collectivization and personalization) the question of the Rights of Man admits of no simple or general answer. But we can at least say that any proposed solution must satisfy the following conditions:

a The individual in a human society in process of collective organization has not the right to remain inactive, that is to say, not to seek to develop himself to his fullest extent: because upon his individual perfection depends the perfection of all his fellows.

b Society, embracing the individuals which comprise it, must in its own interest be so constituted that it *tends* to create the most favorable environment for the full development (physical and spiritual) of what is special to each of them. A commonplace indeed: but one where it is impossible to lay down rules for particular cases, since they vary according to the level of education and the progressive value of the diverse elements to be organized.

c Whatever measures may be adopted to this end, there is one major principle which must be affirmed and always upheld: in no circumstances, and for no reason, must the forces of collectivity compel the individual to deform or falsify himself (by accepting as true what he sees to be false, for example, which is to lie to himself). Every limitation imposed on the autonomy of the element by the power of the group must, if it is to be justified, operate only in conformity with the free internal structure of the element. Otherwise a fundamental disharmony will arise in the very heart of the collective human organism.

Three principles therefore:

The absolute duty of the individual to develop his own personality.

The relative right of the individual to be placed in circumstances as favorable as possible to his personal development.

The absolute right of the individual, within the social organism, not to be deformed by external coercion but inwardly superorganized by persuasion, that is to say, in conformity with his personal endowments and aspirations.

Three principles to be explicitly affirmed and guaranteed in any new Charter of Humanity.

PARIS, MARCH 22, 1947. UNESCO 1949, PP. 88–9.

CHAPTER 13

THE HUMAN REBOUND
OF EVOLUTION AND ITS
CONSEQUENCES

1. Introduction: The Rebounding of Evolution

A YEAR AGO I argued in this journal that, observed in a certain aspect (the truly scientific aspect, in my view), the human social phenomenon
affords evidence that the evolution of Life on
earth, far from having come to a stop, is on the
contrary now entering a new phase.[1]

I maintained that, contrary to the commonly
expressed or tacitly accepted view, the era of active
evolution did not end with the appearance of the
human zoological type: for by virtue of his acquirement of the gift of individual reflection Man
displays the extraordinary quality of being able to
totalize himself collectively upon himself, thus extending on a planetary scale the fundamental vital
process which causes matter, under certain conditions, to organize itself in elements which are ever
more complex physically, and psychologically ever

[1] Cf. chapter 10, The Formation of the Noosphere. *Revue des
Questions Scientifiques,* January 1947.

more centered. Thus (provided always that we accept the organic nature of the social phenomenon) we see being woven around us, beyond any unity hitherto acknowledged or even foreseen by biology, the network and consciousness of a Noosphere.[2]

Following upon this I argued that biological evolution is not only being extended beneath our gaze through the development of the human social group, but that it gives the impression of rebounding upon itself. And indeed although in the prehuman stages of evolution the gradual growth of consciousness in animals (see Section 2, later) does not appear to have had any appreciable effect on the course or speed of their zoological evolution, from the time of Man the evolutionary mechanism undergoes a radical change. For Man, by the act of "noospherically" concentrating himself upon himself, not only becomes reflectively aware of the ontological current on which he is borne, but also gains control of certain of the springs of energy which dictate this advance: above all, collective springs, in so far as he consciously realizes the value, biological efficiency and creative nature of social organization; but also individual springs in as much as, through the collective work of science, he feels himself to be on the verge of acquiring the power of physicochemical control of the operations of heredity and morphogenesis in the depths of his own being. So we may say that since by a sort of chain-reaction consciousness, itself born of complexity, finds itself in a position to bring about "artificially" a further increase of complexity in its material dwelling (thus en-

[2] It should be noted here that by its very nature as a centered, "reflective" collectivity, the Noosphere, while occupying the same spatial dimensions as the Biosphere, differs from it profoundly in its structure and quality of vital completion. Whereas the Biosphere in its essence is complexity linked but divergent and diffused, the Noosphere combines in itself the properties of a planetary zone (or sphere) and those of a sort of higher individuality endowed with something in the nature of a superconsciousness.

gendering or liberating a further growth of reflective consciousness, and so on . . .) the terrestrial evolution of Life, following its
main axis of hominization, is not only completely altering the scale
of its creations but is also entering an "explosive" phase of an entirely new kind.

To me this appears the most satisfactory interpretation of the
present state of Life on the surface of the earth; despite a regret-
table recrudescence of racialism and nationalism which, impressive though it may be, and disastrous in its effect upon our private
postwar lives, seems to have no scientific importance in the overall
process: for the reason that any human tendency to fragmentation,
regardless of its extent and origin, is clearly of an *order of magnitude*
inferior to the planetary forces (geographic, demographic, economic and psychic) whose constantly and naturally growing pressure must sooner or later compel us willy-nilly to unite in some
form of human whole organized on the basis of human solidarity.[3]

I shall not here attempt the perilous and fruitless task of prognosticating the stages, or the probable duration, or the terminal
modalities of this inevitable unification of the human species. I will
only recall that, by virtue of its convergent nature, hominization is
scarcely conceivable (seen from the point at which we find ourselves)
except as terminating, whatever road it follows, in a *point of collective
reflexion* where Mankind, having achieved within and around itself,
technically and intellectually, the greatest possible coherence, will
find itself raised to a higher critical point—one of instability, tension, interpenetration and metamorphosis—coinciding, it would
seem, with what for us are the phenomenal limits of the world.

But I wish, on the other hand, to insist upon certain consequences, of an immediately practical kind, ensuing from what I

[3] See later, under "Conclusion," remarks on "the critical lines of attraction" between human particles.

have called the "reflective rebound" of evolution upon itself; consequences which all converge in a single generalized phenomenon—namely, a certain irresistible functional incorporation of the psychic within the physicochemical which occurs in the process of evolution from the time of the coming of Man.

Let me explain.

2. Emergence of Purposive Thinking

FROM THE EARLY beginnings of biological evolutionary theory, in the nineteenth century, two trends of thought have prevailed in scientific circles, developing side by side without mingling to any appreciable extent. No one doubts any longer that the world of living forms is the outcome of increasingly complex associations between the material particles of which the universe is composed.[4] But how are we to envisage the generative mechanism of this "complexification"? It is very certain that matter on Earth is involved in a process which causes it to *arrange itself*, starting with relatively simple elements, in ever larger and more complex units. But how are we to account for the origin and growth of this process of arrangement? Does it proceed from within, being conceived and developed further by psychic forces analogous to our human power of invention? Or does it simply come from outside, through the *automatic* selection of the more stable (or progressive) groupings among the immense number of combinations fortuitously and incessantly produced in Nature? It is curious to note how since the time of Lamarck and Darwin these two theories, while deepening in their respective ways, have become more sharply opposed. And with varying fortunes.

---

[4] cf. *Les Études*, May 1946—here pages 103ff.

Neo-Darwinism at present holds the ascendancy in the eyes of biologists, partly owing to a clearer and more statistically substantiated definition of "the fittest," but principally because of the immense part, now recognized by modern genetics, played by the "action of large numbers" in the formation of species.

It is to this conflict of opinion—so apparently unyielding that one is inclined to wonder if it has not escaped from the realm of fact to become a simple clash of metaphysical or temperamental preferences—that the hypothesis of a human rebounding of Evolution does, I believe, if Science will accept it, bring a solution and a satisfactory issue. And in the following manner.

That Man displays powers of invention in the creative use of his reflective faculties, that is to say, acts in accordance with an inner sense of purpose, is so apparent that no one has ever thought of denying it. But this fact remained suspended in a void, and without precise significance, while Man and his activities appeared to be isolated and as it were unattached in the bosom of Nature. The whole situation changes if, for reasons solidly bound up with the general structure of the Universe, we regard the process of hominization, with all its accoutrement of social and "artificial" arrangements, as a prolongation and organic continuance of the grand cosmic phenomenon of the vitalization of matter. It then appears that if the neo-Darwinians are right (as they possibly and indeed probably are) in claiming that in the prehuman zones of Life there is nothing but the play of chance arrangement or selection to be detected in the advance of the organized world, from the time of Man, on the contrary, it is the neo-Lamarckians who have the better of the argument, since at this level the forces of internal arrangement begin to be clearly manifest in the process of evolution. Which amounts to saying that biological purposiveness (as with so many other physical parameters of the universe) is not everywhere apparent in the living world, but that it only shows itself above a certain level—its appearance coinciding, not with any particular stage between the Immense and the Infinitesimal but (as in the case of Life itself) with the attainment of a certain value in the "axis of complexities." Below this critical point everything happens (perhaps?) as though the rise of Life were automatic. But above it the forces of free choice and inner direction come to light, and from this moment it is they that tend to *take charge.*

The point I wish to make is this. In the present state of hominization, as we see it in progress today, the statistical influence of chance and the part played by natural selection certainly continue to be enormous. Compared with this immense passive field (the Darwinian) it may seem that the (Lamarckian) ground gained by our inventive efforts amounts to very little. But let us make no mistake about it. However minute the bud may be, however small the seed, it is precisely here that the power of renewal and rebounding of the living world is concentrated. Born under the appearance and the sign of Chance, it is only through reflective purposiveness, slowly acquired, that Life can henceforth hope to raise itself yet higher, by autoevolution, in the twofold direction of greater complexity and fuller consciousness. Indeed, from now on all the hopes and future of the Universe are dependent on the propitious and stubborn working of this scarcely born power of internal "self-arrangement."[5]

And this means, if we are not to regard the world as having become suddenly meaningless and contradictory, that we are entitled to attribute the value of experimental and physical reality to everything, within us and around us, which shows itself to be a *necessary*

---

[5] Teilhard uses the English word.

condition for the preservation and heightening in Man of his powers of invention and purposive thinking.

3. The Control and Preservation of Purposive Thinking

AFTER A SHORT period of untroubled proprietorship every new
source of energy, as we know by experience, gives rise to two related problems, that of the limitations to be imposed on it and
that of its preservation. The new force must not be allowed to get
out of hand or to exhaust itself. The same applies (although we
have thought less about it) to the source of energy abruptly released by Nature through Man which I have called the "force of
purposive thinking." In its early forms human inventive power, as
we still see in children, may be likened to a game. In those first
manifestations of the power of reflective arrangement, everything
appears simple, harmless and even beneficent, giving no hint of a
moment to come when we can no longer go on playing. But as the
phenomenon spreads and develops within a Mankind in process
of becoming adult, what once looked like a game is suddenly
found to be deadly earnest. On the one hand the "sorcerer's ap-
prentice" by dint of fumbling has laid hands on forces of such
power that he begins to be afraid of causing some disaster in Nature. And on the other, finding that by his discoveries he has acquired certain keys to the mastery of the world, he begins to
realize that if he is to be equal to the situation he is bound, in his
role of "quasi-demiurge," to establish principles and a faith regarding the future and the value of the task that is henceforth imposed upon him.

Two roads, as I shall seek to show, by which certain energies
and certain radiations, moral and mystical in their nature, in-

evitably make their appearance at the heart of the biological flux of evolution.

a The Moral Ordering of Invention. By "invention" I mean to designate, in the widest sense of the word, everything in human activity which in one way or another contributes to the organico-social construction of the Noosphere and the development within it of new powers for the arrangement of matter. From the "materialist" point of view the progress of invention in this sense will be entirely governed by the pressure of external necessities, primarily economic. But it has become plain (in particular since the last war) that however urgent may be the planetary pressures driving us to unite, they cannot operate effectively in the long run except under certain psychic conditions, some of which arise out of the human neomystique to be discussed in the next paragraph, but the rest of which merely recall and reexpress, with a precise biological foundation, the broad lines of the empirical and traditional Ethics which has been evolved in some ten millennia of civilization. It is enough for me to cite the twofold respect for things and for personality in the individual. Clearly whatever we may seek to build will crumble and turn to dust if the workmen are without conscience and professional integrity.[6] And it is even more abundantly clear that the greater our power of manipulating inert and living matter, the greater proportionately must be our anxiety not to falsify or outrage any part of the reflective consciousness that surrounds us. Within a short space of time, owing to the acceleration

---

[6] In this connection it is interesting to note the extent to which the lie (a relatively minor evil in more restricted groups) is fast becoming an inhibiting major vice in large social organisms, so that one might say that (like hatred—and the *taedium vitae*) it tends to constitute a major obstacle to the formation of a Noosphere.

of social and scientific developments, this twofold necessity has become so clearly urgent that to refer to it is to utter a commonplace. In recent years voices of alarm have been raised periodically in many quarters pointing to the fast-growing gulf between technical and moral progress in the world today. The perils of the situation are plain to everyone. But do we not underestimate and misunderstand its deep significance?

Many people, I am convinced, still regard the higher morality which they look for and advocate as no more than a sort of compensation or *external* counterbalance, to be applied to the human machine from outside in order to adroitly offset the overflow of Matter within it. But to me the phenomenon seems to display much more intrinsic and fundamental harmony and much closer affiliations. The ethical principles which hitherto we have regarded as an appendage, superimposed more or less by our own free will upon the laws of biology, are now showing themselves—not metaphorically but literally—to be a condition of *survival* for the human race. In other words Evolution in *rebounding* reflectively upon itself, acquires *morality* for the purpose of its further advance. In yet other terms, and whatever anyone may say, above a certain level, technical progress necessarily and functionally adds moral progress to itself. All this is surely proof that the two events are interdependent. In fact, the pursuit of human knowledge cannot be carried in concrete terms beyond a certain stage without this power of reflective arrangement becoming automatically charged with internal obligations which curb and direct it; while at the same time, as we shall see, it engenders around itself an entirely new atmosphere of spiritual needs.

b *The Spiritual Nourishment of Human Endeavor.* It is surprising to note, among the increasingly numerous theorists who, under the pressure of events, are beginning to speculate on the future of the phenomenon of man, a sort of tacit agreement whereby vital energy is treated as though it were a constant, both in quality and quantity, like solar radiation or the force of gravity. This postulate of invariability seems at first sight to be admissible in the "Darwinian" zones of Life, where the instinct of self-preservation predominates (this seeming by its nature to be more or less constant among organized beings), but it certainly loses all value in the "Lamarckian" or human zone, where biological evolution, from being passive, becomes active in the pursuit of its purpose. As we know very well in ourselves, and as every leader of men has discovered, human creative energy, according to the degree of temperature generated within it (on a scale, that is to say, between enthusiasm and revulsion) can in a matter of instants jump "from plus to minus infinity."

If, therefore, we accept the idea of a reflective rebounding of evolution, it is not enough to reckon the future of the world in terms of reserves of mechanical energy and food supplies, or the probable longevity of the earth. As I have said elsewhere,[7] the evolutionary vigor of Mankind can wither away although it be surrounded by mountains of coal, oceans of petroleum and limitless stocks of wheat; it can do so as surely as in a desert of ice, if Man should lose his impulse, or worse, develop a distaste for ever-increased growth "in complexity and consciousness." With all respect to the materialist school, which still refuses to examine *human* biology, it is undeniable that in Man the external drive of Life tends to be transformed and turn inward to become an *ardor* for Life. Try to get productive work out of a workman, an engineer or a scientist who is "pissed off"! So in the first place, if Evolution is to continue, it is this impetus which must be maintained in the heart of Man and encouraged

---

[7] *L'Énergie Humaine,* 1937.

to grow at all costs. Failing that upward current, almost nothing will move; whereas with it, everything will happen almost of its own accord in the higher zones, those that are truly progressive, of invention and discovery. But how are we to tap this deep, primordial well?

If the problem of sustaining this human impetus does not trouble the theorists I have referred to, it is, I suppose, because they assume that cases of revulsion will be as exceptional in the future as they have been in the past—that a sufficient degree of physical health or euphoria will maintain vital pressure at a positive level, moderate but adequate, within the human mass. But is not this to beg the whole question? Not only have powers of reflection and invention been added to Life through hominization, but so has the formidable endowment of criticism. However exuberant our vitality, however rich and sanguine our temperament, it is already becoming impossible, and must inevitably become more so, for us to give ourselves wholly to any creative undertaking if we cannot justify it in rational terms. That is why if Man at this moment finds himself faced with the burden not merely of submitting to the evolutionary process but of consciously furthering it, we may be sure that he will seek, and rightly, to avoid the responsibility and pangs which this entails if the objective does not seem to be *worth the effort*. Which amounts to saying that the Universe, of psychic or psychological necessity (here they come to the same thing) must possess properties fulfilling the functional needs of reflective action. Otherwise apathy and even disgust will pervade the human mass, neutralizing or reversing every vigorous impulse at the heart of Life.

What and how many are these basic properties, these *sine qua non* conditions, which we are bound to postulate and presume to be incorporated in the structure of the surrounding world if Evolution, henceforth hominized, is to continue?

In our present state (or more exactly, stage) of psychic awareness it seems to me that they can be brought down to two, very closely related.

The first, as I have argued at length in my chapter on the Noosphere, is that in one way or another Consciousness, the flowering of Complexity, must survive the ultimate dissolution from which nothing can save the corporeal and planetary stem which bears it. From the moment when Evolution begins to *think itself* it can no longer live with or further itself except by knowing itself to be *irreversible*—that is to say, immortal. For what point can there be in living with eyes fixed constantly and laboriously upon the future, if this future, even though it take the form of a Noosphere, must finally become a zero? Better surely to give up and die at once. In terms of this Absolute it is sacrifice, not egotism, that becomes odious and absurd. Irreversibility, then, is the first condition.

The second condition, no more than an amplification of the first, is that the irreversibility, thus revealed and accepted, must apply not to any one part, but to all that is deepest, most precious and most incommunicable in our consciousness. So that the process of vitalization in which we are engaged may be defined at its upper limit (whether we envisage the system as a whole or the destiny of each separate element within it) in terms of "ultrapersonalization." The necessity of this must be stressed, since the degree of personalization (or "centration," which comes to the same thing) of a cosmic element being finally the sole parameter by which we can measure its absolute biological value, a world presumed to be heading toward the Impersonal (the word being interpreted in its normal sense of "infrapersonal") becomes both unthinkable and unliveable.

An *irreversible* rise toward the *personal*: unless it satisfies one or other of these two conjoined attributes, the Universe (psychoanalytically dosed, if I may put it that way) can only become stifling

for all reflective activity, that is to say, radically unsuited to any rebound of Evolution. But we are agreed that such a rebound is
preparing and indeed has already begun. So we must conclude,
unless we favor the idea of a world destined to miscarry through a
fault in its construction, that evolutionary irreversibility and personalization (despite their implied anticipation of the future) are
realities not of a *metaphysical* but of a physical order, in the sense
that, like the dimensions of Time and Space, they represent general conditions to which the totality of our proceedings must conform.

Failing these conditions, as I have said, everything at the level
of Man will cease to move. On the other hand it seems to me that,
provided they are fulfilled, nothing can seriously interfere with our
natural taste, our impulse, that is to say, toward invention and research. The world will have become habitable for Thought. But is
it enough for the world as we are now picturing it to be simply liveable, capable, all things considered, of fostering some degree of
taste for life? Must it not rather be wholly *delectable*, if it is to be
wholly consistent with itself?

Strange though it may seem, we are here confronted, if we
seek to define our Universe in relation to other imaginable kinds
of universe, with the necessity and importance of determining
what may be called its "coefficient of activation," that is to say,
the degree in which it possesses the quality of stimulating the centers of reflective activity contained within it. Theoretically, in
virtue of what I have said, a whole series of activations (provided
they are positive) is conceivable, each in itself sufficing to create a
liveable world. But in practice—does not some sixth sense warn
us of this?—one value alone is admissible in the experienced reality of action, one alone can truly satisfy us: namely, *the greatest of*

all, in relation to what we are. I do not propose to embark upon any analysis or defense of the very particular kind of optimism which does not for a moment claim that we are living in the best of all possible worlds, but only (a quite different matter!) the most "activating." Let me simply observe that here perhaps we have a basis for prognosticating, in the broadest outline, the religious evolution of the world of tomorrow. From the strictly "noodynamic"[8] viewpoint which I had adopted, it may be said that the historic rivalry of mysticisms and creeds, each striving to conquer the earth, represents nothing but a prolonged groping of the human soul in search of a conception of the world in which it will feel itself to be more sensitized, more free and active. This surely means that the faith which finally triumphs must be the one which shows itself to be more capable than any other of inspiring Man to action. And it is here, irrespective of all philosophical or theological considerations, that Christianity decisively takes the lead with its extraordinary power of immortalizing and personalizing in Christ, to the extent of making it lovable, the time-space totality of Evolution.

4. Conclusion: Something New under the Sun

IN SHORT, AS I said at the beginning, the terrestrial evolution of Life, if it is really to continue as hominization extended to the scale of the Noosphere, cannot rebound in a new spring forward without acquiring a morality, and, to the extent that it needs a "faith,"

---

[8] "Noodynamic": the dynamic of spiritual energy, dynamic of the Spirit. I have ventured to use this neologism because it is clear, expressive and convenient; also because it affirms the necessity for incorporating human psychism, Thought, in a true "physics" of the World.

without becoming "mysticized." Which amounts to saying that the complexification of Matter, at the point it has now reached in the human social organism, is physically incapable of advancing further if the Mind does not play a part, not only with its capacity for technical organization, but with its purposive and affective powers of arrangement and inner tension. Which again amounts to saying in another way that from the time of Man (above all, modern Man) the factor *consciousness*, which for a long time perhaps represented no more than a secondary and accessory effect in Nature, a simple superstructure of the factor *complexity*, is finally becoming individualized in the form of an autonomous spiritual principle. For its reflective and inventive forward spring it is in some sort necessary that Life, duplicating its evolutionary motive center, should henceforth be sustained by *two* centers of action, separate and conjoined, one of consciousness and the other of complexity. And herein, if I am right, we may find a bridge of an experimental kind flung across a gap which has hitherto been held to be scientifically unbridgeable. In hominized evolution the Physical and the Psychic, the Without and the Within, Matter and Consciousness, are all found to be functionally linked in one tangible process. Setting aside all metaphysics, the two terms in each of these pairs are *articulated* in a quasi-measurable fashion one with the other; with the twofold result not only of at last affording us a unified concept of the Universe, but also of breaking down the two barriers behind which Man was coming to believe himself to be for ever imprisoned—the magic circle of phenomenalism and the infernal circle of egocentrism.

Gone, first of all, the magic circle of phenomenalism, which, we have been assured, must inexorably restrict our gaze to a limited horizon beyond which lies not merely the unknown but the absolutely unknowable. How much has been said even recently[9] about our powerlessness to penetrate in this sense beyond the primitive vision shared by the earliest human minds; that is to say, the impossibility of our advancing a step toward the direct or indirect perception of all that is hidden behind the veil of tangible experience! But it is just this supposedly impenetrable envelope of pure "phenomenon" which the rebounding thrust of human evolution pierces, at least at one point, since by its nature it is irreversible. This does not mean that we can see what lies beyond and behind that transphenomenal zone of which we now have an inkling, any more than, having discerned the shape of the earth, we can foresee the landscape lying below the horizon. But at least we know that something exists beyond the circle which restricts our view, something into which we shall eventually emerge. It is enough to ensure that we no longer feel imprisoned.

Gone, too, (at least virtually and in aspiration), is the infernal circle of egocentrism, meaning the isolation, in some sort ontological, which prohibits our escape from self to share the point of view even of those we love best: as though the Universe were composed of as many fragmentary universes, *repelling* each other, as the sum total of the centers of consciousness which it embraces. Who can measure the long chain of harmful, closely interlinked effects which this elemental separatism automatically creates and fosters, by an effect of mass and resonance, within the process of totalization now taking place in Mankind? The iron laws binding eco-

---

[9] "It is perhaps inevitable that, having reached the limit beyond which the sure basis of experience fails us, the human spirit, in its impotence, can do no more than revolve in the magic circle of traditional interpretations." (Betti, Director of the Chemical Institute of Bologna.) "When it comes to those questions on the border-line of the unknowable, all the accumulated knowledge of twenty-five centuries has done no more than feed the argument, without advancing us a single step toward the solution." (Tannery, *Pour la Science Hellène*; quoted by J. Benda in *La Tradition de l'Existentialisme*.)

nomic factors, the irrepressible recurrence of nationalisms, the apparent inevitability of war, the insoluble Hegelian conflicts "of master and slave"; what are these supposedly unalterable necessities of the human condition, except, finally, the diverse expression and outcome of exteriority and a mutual antagonism between the individual seeds of thought which we are? . . . Here again let us throw back our heads and breathe freely! For if it be true that the tide of evolutionary totalization sweeping us along requires, for its viability, not only that we must progress toward some form of irreversible unity, but also that this progress must be in the personal sphere, is not this a positive reason for believing that sooner or later something must happen in the world whereby certain basic conditions of the human phenomenon will undergo modification? If our "person" is not to be lost in the vast plurality of Mankind within which it is gradually, and of inescapable physical necessity, becoming integrated; if totalization is to set us free instead of simply mechanizing us; then we must look for and allow for a change of regime. We must assume that under the rapidly mounting pressures forcing them upon one another the human molecules will ultimately succeed in finding their way through the critical barrier of mutual repulsion to enter the inner zone of attractive.[10]

From that point on we shall be entering a new world of relationships where the hitherto impossible may become simple, being enacted in other dimensions and another environment.

If such a vista still seems to us fantastic, it is simply that we lack imagination. But scientific reason is there to sustain and guide us. Some hundreds of thousands of years ago, upon the first emergence of reflective consciousness, the Universe was surely and beyond question transformed in the very laws of its internal devel-

[10] This is an old idea which I advanced nearly twenty years ago in an unpublished essay entitled *The Spirit of the Earth*. (Now printed in *L'Energie Humaine*, pp. 25–57.)

opment. Why, then, should we suppose that nothing entirely new will appear under the sun of tomorrow, when the rebounding of Evolution is in full flood?

To sum up.

If social totalization and scientific technology are regarded as they should be, as constituting a direct prolongation, in a human context, of the grand process of the vitalization of Matter, it follows that, from the coming of Man, biological evolution not only rebounds (on a new scale and with new resources) but that it rebounds reflectively upon itself. The Darwinian era of survival by Natural Selection (the vital thrust) is thus succeeded by a Lamarckian era of Super-Life brought about by calculated invention (the vital impulse). In Man evolution is interiorized and made purposeful; and at the same time, in the degree in which the strivings of human inventiveness need to be controlled in their operation and sustained in their energies, it imposes upon itself a moral order and "mysticizes" itself. *In abstracto* or *in individuo*, technical achievement and moral virtue, science and faith (faith in the future) may seem to be things that are not only separate but even opposed to one another. But *in the concrete reality of Total Evolution*, and beyond a certain degree of Complexity and Consciousness, each of necessity requires the other, because Matter, once hominized, can positively not continue the superarrangement of itself upon itself except in a specific psychic atmosphere.

Thus a precise functional interlocking of physical and spiritual energy may be discerned. And thus is revealed the necessity for the Universe to present itself to our experience as an irreversible medium of personalization, if the human rebounding of Evolution is not to be stifled at birth.

SAINT-GERMAIN-EN-LAYE, 23 SEPTEMBER, 1947.

REVUE DES QUESTIONS SCIENTIFIQUES, APRIL 20, 1948.

CHAPTER 14

TURMOIL OR
GENESIS?

The Position of Man in Nature and the
Significance of Human Socialization

Is there in the Universe a Main Axis of Evolution?
(An attempt to see clearly)

Introduction

NO ONE CAN any longer doubt that the Universe, conceived in experimental or phenomenal terms, is a vast temporo-spatial system, corpuscular in nature, from which we cannot sensorially escape (even in thought) either backward or forward or by circumventing it. Viewed in this light everything in the world appears and exists as a function of the whole. This is the broadest, deepest and most unassailable meaning of the idea of Evolution.

But it raises a question. How are we to envisage the operation of such a system, which by its nature is both organic and atomic? Is its movement one of disorderly or controlled impulses? Is the world amorphous in structure, or does it on the contrary show signs of containing within itself a favored axis of evolution?

Following the principle that the greater coherence is an infallible sign of the greater truth, I propose to demonstrate that such an axis does in fact exist, and that it may be defined in terms of the following three (or even four) successive propositions or approximations, each of which clarifies and substantiates the one preceding it on a single line of experience and thought:

a Life is not an accident in the Material Universe, but the essence of the phenomenon.

b Reflection (that is to say, Man) is not an incident in the biological world, but a higher form of Life.

c In the human world the social phenomenon is not a superficial arrangement, but denotes an essential advance of Reflection.

To which may be added, from the Christian point of view:

d The Christian phylum is not an accessory or divergent shoot in the human social organism, but constitutes the axis itself of socialization.

Let us look in turn at the separate links in this chain of propositions, each of which, as we shall see, represents a test whereby we may better know and measure, in the scale of spiritual values, the worth and position of others.

Proposition I. Life Is Not an Epiphenomenon in the Material Universe,
but the Central Phenomenon of Evolution

PLACED AND OBSERVED within the general framework of Matter as it is now revealed to science, Life may seem to be of tragically little importance in the Universe. In spatial terms we know for certain of its existence only on an infinitesimally small body in the solar system. In terms of Time its whole planetary duration is no more than a flash in the huge course of sidereal development. And structurally its extreme fragility seems to relegate it to the humblest and lowest place among all the substances engendered during the physicochemical evolution of cosmic matter. We can hardly wonder, in the circumstances, that agnostics such as Sir James Jeans and Marcel Boll, and even convinced believers like Guardini, have uttered expressions of amazement (tinged with heroic pessimism or triumphant detachment) at the apparent insignificance of the phenomenon of Life in terms of the cosmos—a little mold on a grain of dust . . .

Small wonder, I repeat: but it is no less astonishing that minds so outstanding should not have perceived the possibility and the advantages of adopting a precisely opposite viewpoint. Life seems to occupy so small a place in Space-Time, that it cannot reasonably be regarded as anything other than incidental and accidental. That is the difficulty. But why should we not reverse the position and say: "The fact that Life is so rarely encountered in the sidereal immensity is precisely because, representing a higher form of cosmic evolution, it can only come into existence in privileged circumstances of time and place." We shall see the full force of this argument (based on the premise that Life, everywhere and for ever struggling to assert itself, is liable to appear at any point in the Universe when the conditions are favorable) only when we come to the end of this paper: when, that is to say, we have perceived the full coherence and fruitfulness of the mental and moral attitudes to which it gives access. But it is important to insist at the outset on the fundamental point that (despite all contrary appearances and prejudices) the best way of scientifically explaining the World is to make up our minds to regard animate beings, not as a fortuitous by-product but as the characteristic and specific higher aim of the universal phenomenon of Evolution.

Let us strip Life of all its anatomical and physiological superstructure, bringing it down to the essentials of its physicochemical nature. Reduced to its basic mechanism it shows itself to be a straightforward process of increasing complication whereby Matter contrives to arrange itself in particles of ever greater volume, ever more highly organized. But do we not find that at the same time its seeming weaknesses, its fragility and appearance of extreme localization in time-space, tend to vanish? For underlying these supposedly "exceptional" cellular arrangements we have first the far vaster world of molecules, and underlying this again the immense and decidedly cosmic world of atoms; two worlds displaying, the first by its interatomic arrangements and the second by its nuclear groupings, (each in its own way and through different procedures) precisely the same tendency to "fall" into increasingly organized states of complexity.[1] Thus considered, the era of the Organic (living) which may have appeared so exceptional in Nature becomes no more than a further instance, at a particularly

---

[1] We may seek to distinguish the phases or pulsations of the cosmic rise into complexity (that is to say, into the Improbable) as follows:

   a The preatomic phase: formation of nuclei and electrons;

   b The atomic phase: grouping of nuclei in atoms (fixed and limited number of free "compartments");

   c The molecular phase: grouping of atoms in finite or indefinite chains;

   d The cellular phase: grouping of molecules in centered clusters.

   In all these cases, up to but excluding Man, the arrangement seems to have been brought about mainly by the working of chance and of grouping; but in phases *a* to *c* the majority of the groups (except in the case of very large molecules) represent knots of *stability*, whereas in phase *d* the arrangements that survive represent privileged centers of *activity*.

high level, of the operation of the same law that governs the whole of the Inorganic. So finally we find the Universe from top to bottom brought within a single, immense coiling movement[2] successively generating nuclei, atoms, molecules, cells and metazoa—the special properties of Life being due solely to the extreme (virtually *infinite*) degree of complexity attained at its level.

Thus the World falls into order, it organizes itself, around Life, which is no longer to be regarded as an anomaly but accepted as pointing the direction of its advance (evidence in itself that the axis was well chosen!). And what is more up to a point its progress becomes *measurable*: for, as observation shows, it is the nature of Matter, when raised corpuscularly to a very high degree of complexity, to become centered and interiorized—that is to say, to endow itself with Consciousness. This means that the degree of consciousness attained by living creatures (from the moment, naturally, when it becomes discernible) may be used as a parameter to estimate the direction and speed of Evolution (that is to say, of the Cosmic Coiling) in terms of absolute values.

Let us adopt this method and see where it leads us.

Proposition II. Human Reflection Is Not an Epiphenomenon of the Organic World, but the Central Phenomenon of Vitalization

IF, AS WE have agreed, Life is the spearhead of Evolution, does Life in its turn afford us a pointer to the direction of its advance?

---

[2] It may be as well here to distinguish between the two types of "coiling" or "in-folding" in the evolutionary process.

a The Coiling of Mass, which subdivides Matter without organizing it (e.g., the stellar masses);

b The Coiling of Complexity, which organizes elementary masses in ever more elaborate structures.

This again is an idea that the latest scientific research does not seem to favor at first glance. Just as Life itself seems to fade into intangibility and insignificance within the sidereal immensities known to astronomy, so does the happy simplicity which seemed to indicate a gradual and steady rise of consciousness from the lower animals to Man lose distinctness in the extraordinary diversity and profusion of living forms now known to biology. Formerly "instinct" could be treated as a sort of homogeneous quantity varying (something like temperature) on a scale running from zero to the point of Reflection representing human thought. Now we have to accustom ourselves to seeing things differently. It is not along a single line that Consciousness has emerged and is increasing on earth, but along an immense fan of nervures, each nervure representing a particular kind of sensory perception and knowledge. There are as many wavelengths of consciousness as there are living forms.[3] How can we venture to assert that in this spectrum or spreading sheaf of psychisms, any *single* line can exist? Hence the reluctance of many biologists to fix upon a scale of values for use within the animal kingdom. Is Man really more than a protozoan? It has been possible for the question to be seriously asked and left unanswered. But if there were really no answer we should be obliged to conclude that, although the course of Evolution was "directed" up to the emergence of Life, beyond that point all that goes on is a scat-

All Mass-Coilings certainly do not result in Coilings of Complexity; but on the other hand all Coilings of Complexity seem to originate or be conditional upon a Mass-Coiling—for example, Life, which could only be achieved on the physical foundation of a planet.

[3] i.e., in seeking to grasp the interior world and associative faculties of an animal it is not enough to try to diminish or decenter our own picture of the world: we have to modify our angle of vision and our way of seeing. Failing this we fall into the anthropomorphic illusions which cause us to be amazed at the phenomena of mimetism, or by mechanical arrangements which we ourselves could only carry out with the full aid of science, whereas the insect or the bat seems to have acquired the skill directly.

tering in every direction. We are left with no trail to follow unless
we decide, for sufficient reasons, to attribute a unique and privileged value to *reflective* consciousness.

It has become the rather unconsidered fashion, since Bergson,
to decry intelligence as compared with other forms or aspects of
cognition. To the extent that this is simply a reaction against a
static and abstract rationalism, it is wholly salutary; but it becomes
pernicious if it goes so far as to cause us to overlook what is truly
exceptional and essential in the phenomenon of Thought—the
power of Consciousness to center so perfectly upon itself as to be
able to situate itself (itself and the Universe at the same time) in the
explicit framework of a present, a past and a *future*—that is to say,
in a Space-Time continuum. The more we reflect upon the revolutionary consequences ensuing from this transformation of the
laws hitherto governing the world—the growth of powers of *foresight and invention*, prompting and guiding a "planned" rebound of
Evolution!—the more must we be persuaded that to regard Intelligence as an anomaly and even a disease of Consciousness is as
absurd and sterile as to regard Life as a mold on the earth's surface; and the more do we find ourselves drawn toward another interpretation of the facts, which may be expressed as follows: It is
perfectly possible that in the general spectrum of Life the line ending in Man was originally no more than one psychic radiation
among countless others. But it happened, for some reason of hazard, position or structure, that this sole ray (this is an experiential
fact) among the millions contrived to pass the critical barrier separating the Unreflective from the Reflective—that is to say, to enter
the sphere of intelligence, foresight and freedom of action. Because it did so (and although in a sense, I must repeat, this ray was
only one attempt among many) the whole essential stream of terrestrial biological evolution is now flowing through the breach
which has been made. The cosmic tide may at one time have

seemed to be immobilized, lost in the vast reservoir of living forms; but through the ages the level of consciousness was steadily rising behind the barrier, until finally, by means of the human brain (the most "centro-complex" organism yet achieved to our knowledge in the universe) there has occurred, at a first ending of time, the breaking of the dykes, followed by what is now in progress, the flooding of Thought over the entire surface of the biosphere.

Thus regarded, everything in the history of the world takes shape, and what is better, everything *goes on.*

Proposition III. Socialization Is Not an Epiphenomenon in the
Sphere of Reflective Life but the Essential Phenomenon of Hominization

I BELIEVE THAT few readers will quarrel with my reasoning in favor of Propositions I and II. Where that part of the argument is concerned, the way through the jungle of facts has been cleared by a century of research and discussion. We may assert today that there is almost complete unanimity among scientists regarding the central position of Life in the Universe, and of Man in Life. It is beyond this point—beyond Man in his anatomical and spiritual individuality—that the path vanishes in the undergrowth and the dispute begins. We have now entered the battle: let us see what the position is.

What hinders and even prevents us from advancing beyond this point is our evident inability to conceive of anything more organically complex or psychically centered than the human type emerging in Nature as it now is. Hence the instinctive tendency, so widespread even among men of science, to regard the tide of Life on earth as having for practical purposes ceased to flow. According to this view, Life, having reached the reflective stage, must not only disperse in diverging ethnocultural units, but must finally culminate (and one might say, *evaporate*) in separate individualities, each within the enclosed sphere of its sensibilities and knowledge representing an independent, absolute summit of the Universe.

That is one way of looking at it. But before we acquiesce in a solution which to me seems nothing but the implied admission of a dead end, we need to be quite sure that the forces of vitalization really do possess no outlet upon earth, above the level of the human individual. We are told that the way ahead is completely closed. But have those who believe this given any thought to the forces of socialization?

From habit, and from ignorance, we are inclined to consider the human social phenomenon as no less commonplace and uninteresting than the human phenomenon of reflection. What, we ask in effect, can be more sadly natural than that the human particles, since, unluckily for them, they gather in crowds and masses, should feel the need to organize themselves so as to make existence tolerable? What is this but a process of necessary adjustment, with no mystery about it? That is the view taken by many people as they gaze with melancholy disquiet at the turbulent swell of humanity; and by it the whole edifice of human relationships and social structures is reduced to the level of a regulated epiphenomenon, having no value or substance of its own, and therefore no future in its own right.

But here, and for the third time, why should we not adopt a position diametrically opposed to the one which is most familiar and, at first sight, most simple? Why not assume instead that, if it is by reason of the cosmic structure, and not by chance, that man is born "legion," by the same token it is not through chance, but through the prolonged effect of "cosmic coiling," that the human layer is weaving and folding in upon itself in the way we see it to be doing? On this basis the fundamental evolutionary process of the Universe does not stop at the elemental level of the human brain and human reflection. On the contrary, at this stage the "complexity-consciousness" mechanism gains an added impulse, acquiring a new dimension through new procedures. It is no longer simply a matter of cells organized by the hazards of natural selection, but of completed zoological units inventively building themselves into organisms on a planetary scale. Adopting this organic view of the social phenomenon, we find that not only does the structure of our terrestrial society become meaningful both in a general sense (the gradual rise of tension or psychic temperature under technico-social pressure) and in detail (the "anatomy" and "physiology" of the Noosphere) but the whole process takes on a *convergent* aspect: the phenomenon of man, seen in its entirety, appears to flow toward a critical point of maturation, (and perhaps even of psychic withdrawal)[4] corresponding to the concentration of collective Reflection at a single center embracing all the individual units of reflection upon Earth.

Further than this we cannot see and our argument must cease—except, as I have now to show, in the case of the Christian, who, drawing upon an added source of knowledge, may advance yet another step.

Proposition IV. The Church Is Neither an Epi- Nor a
Paraphenomenon in the Growth of the Human Social Organism,
but Constitutes the Very Axis (or Nucleus) about Which It Forms

TO THOSE ACCUSTOMED to see in the phenomenon religion nothing more than a purely conventional association of minds in the sphere of the "imaginary," this fourth and final theorem will

---

[4] Necessitated, it would seem, by the requirement of irreversibility developed on the way by the coiling of the Cosmos upon itself.

seem astonishing and may even come under suspicion as "illumi-
nism." Yet it arises directly out of the juxtaposition of two concepts of the World: the one which practical considerations have
just led us to adopt, and the one which every Christian is bound to
accept if he is to remain orthodox. As we know, the belief that the
human individual cannot perfect himself or fully exist except
through the organic unification of all men in God is essential and
fundamental to Christian doctrine.[5] To this mystical superorganism, joined in Grace and charity, we have now added a mysterious
equivalent organism from the domain of biology: the "Noospheric" human unity gradually achieved by the totalizing and
centering effect of Reflection. How can these two superentities, the
one "supernatural," the other natural, fail to come together and
harmonize in Christian thought; the critical point of maturation
envisaged by science being simply the physical condition and experimental aspect of the critical point of the Parousia postulated
and awaited in the name of Revelation? Clearly for the conjunction to be effected it is necessary (as is already happening) for it to
gain possession of many devout minds. But we must be clear that
this change in our vision goes far beyond any purely intellectual
and abstract merging of two complementary pictures, one rational, the other religious, of "the end of the world."

For one thing, by this conjunction Christian cosmology,
harmonized and effectively articulated at its peak with Human
cosmology, shows itself to be fundamentally and in real values
homogeneous with the latter. Thus dogma is no mere flowering of

[5] From the Christian point of view (which in this coincides with the biological
viewpoint logically carried to its extreme) the "gathering together" of the Spirit
gradually accomplished in the course of the "coiling" of the Universe, occurs
in two tempos and by two stages—*a* by slow "evaporation" (individual deaths);
and simultaneously *b* by incorporation in the collective human organism ("the
mystical body") whose maturation will only be complete at the end of Time,
through the Parousia.

the imagination but something authentically born of history; and it is in literal not metaphorical terms that the Christian believer can illumine and further the genesis of the Universe around him in the form of a Christogenesis.

Moreover, by very virtue of the interlocking of the two "geneses" the ascending force of Christianity is directly geared to the propulsive mechanism of human superevolution. To the Christian, for whom the whole process of hominization is merely a paving of the way for the ultimate Parousia, it is above all Christ who invests Himself with the whole reality of the Universe; but at the same time it is the Universe which is illumined with all the warmth and immortality of Christ. So that finally (the point cannot be too strongly stressed) a new impulse becomes possible and is now beginning to take shape in human consciousness. Born of the psychic combination of two kinds of faith—in the transcendent action of a personal God and the innate perfectibility of a World in progress—it is an impulse, (or better, a spirit of love) that is truly evolutionary. We can indeed say of it that it is the only kind of spiritual energy capable of causing the formidable human machine, in which, from what we can see, all the future and all the hopes of Evolution must henceforth be concentrated, to function at full power, without danger from egotism or from mechanization, and to the full extent of its potentialities.

Conclusion

WHAT I SET out to show, and hope to have shown, is that, viewed from a certain angle, the internal stir of the Cosmos no longer appears disorderly: it takes a given direction following a major axis of movement at the completion of which the phenomenon of man becomes detached as the most advanced form of the largest and most characteristic of cosmic processes, that of in-folding. This axis, as we have suggested, may conveniently be determined by means of three successive propositions, so closely linked that we cannot accept any one without being committed to those which precede it, nor, conversely reject any one without being barred from those that follow it.

This coherence (which is not closed, like the coherence of a system, but open like that of a method of, or key to, progressive research)—is so marked that we would have to have extremely grave positive reasons for refusing to face it; and for my part I can see none that is adequate. But it is nevertheless true that, above all if they are taken separately, none of the propositions I have formulated is rigidly deductive or, therefore, conclusive: each is more in the nature of an intuition, that is to say, a kind of choice. So it is possible to part company from the sequence at each stage: but only, it should be noted, if in doing so we accept the alternative choice. But this, to the logical mind, threatens to have dangerous repercussions in the field of action.

As an instance let us take the particularly crucial and meaningful Third Proposition—or option.

Do we accept the idea, strongly supported by fact, that the individual man cannot achieve his wholeness (that is to say, reflect and personalize himself completely upon himself) except in solidarity with all other men, present, past and future? If we do, the awareness aroused in us of being each a responsible element in a rebounding course of Evolution must, at the same time as it gives rise to a desire and reason for action, inspire us with a fundamental sense of obligation and a precise system of moral tendencies. In matters of love or money or liberty, of politics, economics or society, we not only find our main line of conduct and criteria of choice structurally laid down for us ("ever higher in convergence") but furthermore, our instinct for research and creation ("to consummate the Universe in ourselves") discovers endless justification and sustenance. Viewed in this way, everything makes sense, everything glows with life; and the flow of human sap rises to the very heart of the Christian faith.

But if, on the other hand, we refuse to regard human socialization as anything more than a chance arrangement, a *modus vivendi* lacking all power of internal growth, then (excepting, at the most, a few elementary rules safeguarding the living-space of the individual) we find the whole structure of politico-economico-social relations reduced to an arbitrary system of conventional and temporary expedients. Everything in the human world becomes artificial in the worst sense of the word; everything is divested of importance, urgency and interest; Christianity itself becomes no more than a sort of alien proliferation, without analogy or roots in the Phenomenon of Man.

Faced by so wide a divergence of attitudes, can we fail to see that the attempt made in these pages to determine a cosmic axis of evolution, far from being a mere intellectual diversion, is by way of expressing the condition of survival for the human race? And more especially how can we do other than feel that it is about the social phenomenon, according to the degree of central and organic value which we attribute to it, that Mankind is in process of reassessing and regrouping itself?

PARIS, DECEMBER 20, 1947. *L'ANTHROPOLOGIE*, SEPTEMBER 1948.

CHAPTER 15

THE DIRECTIONS
AND CONDITIONS OF
THE FUTURE

THE FUTURE, I mean the human future, certainly contains an element of the unpredictable in itself. Because of the enormous number of physical variables on which it depends, and even more the ever-growing predominance of the psychic (individual choice) over the purely statistical, it seems to be decidedly the case that human evolution goes beyond the bounds of exact calculation. So it would be an error, deserving of vigorous denunciation, to talk as though biology in its forecasts can behave like astronomy. But it is surely no less excessive and dangerous to behave as though our "freedom" were confronted by a future that is completely indeterminate. No matter what the Existentialists may claim, to suppose that the Duration ahead of us resembles a virgin, "isotropic" substance into which we may cut as we please, as expediency dictates and in any direction, is positively and *scientifically* incorrect. Life, and most particularly the extreme point of Life represented by

Mankind, is not simply a *state*. It is on the contrary (I shall come back to this) a vast, directed movement, bound up with the very structure of the Cosmogenesis. It has a "thread" which cannot be suppressed, and which must continue to show itself, in no way impaired, but respected, utilized and expressed, until (and at this point more than ever) it reaches the highest, most conscious forms of its development.

At a time when delegates from all over the world are coming together in a variety of bodies for the purpose of attempting to sketch a first outline of future society, it seems to me essential to set forth the main constructive axes without which it is mere self-delusion to suppose that we can conceive or undertake any ordering or development of the Earth: general *tendencies* of advance and growth, that is to say, which *in certain conditions*—despite our freedom of choice, or better still, because of it—Mankind cannot in any circumstances ignore, and must heed more and more as time goes on.

That is the purpose of this brief note.

1. The General Tendencies

WHATEVER THE PARTICULAR modalities of the form it may eventually take, the world of Man (this is my thesis) already shows certain tendencies in its development, certain lines of embryogenesis, of which we may safely predict that they are definitive and will only become accentuated with time. Without resorting to any systematized theory (I shall propose one later) and simply confining ourselves to the objective study of observable facts, these axes of growth can be reduced to three.

a First—the continuous rise of Social Unification (rise of masses and races). Clearly no one can yet predict the exact nature of the worldgroup toward which events are leading us. But here and now one thing is certain, and it appears to me that its recognition in theory, and acceptance in practice, must be the *sine qua non* of any valid discussion and effective action affecting the political, economic and moral ordering of the present world: this is that nothing, absolutely nothing—we may as well make up our minds to it—can arrest the progress of social Man toward ever greater interdependence and cohesion. The reason is this. The human mass on the restricted surface of the earth, after a period of expansion covering all historic time, is now entering (following an abrupt but not accidental acceleration of its rate of reproduction) a phase of compression which we may seek to control but which there are no grounds for supposing will ever be reversed. What is the automatic reaction of human society to this process of compression? Experience supplies the answer (which theory can easily explain)—*it organizes itself.* To adapt themselves to, and in some sort to escape from, the planetary grip which forces them ever closer together, individuals find themselves compelled (eventually they acquire a taste for it) to arrange their communal lives more adroitly; first in order to preserve, and later to increase their freedom of action. And since the compulsion is applied on a uniform and total scale to the whole mass of humanity the ultimate social organization which it evokes must of necessity be unitary. I have said elsewhere and I repeat it here[1]: it would be easier, at the stage of evolution we have reached, to prevent the earth from revolving than to prevent Mankind from becoming totalized.

b   Second, and correlatively—the growth of generalized technology and mechanization. Here again the facts are clear and the reasons obvious. In a Mankind becoming unified under pressure, its various or-

---

[1] Cf. Chapter 7—Human Planetization.

gans tending in consequence to achieve planetary dimensions, it is inevitable that the mechanical equipment of society will become all-pervading and enormous. But this change of scale alone is not enough in itself to explain the sudden and irreversible rise of the industrial phenomenon which we see taking place around us. What has really let loose the Machine in the world, and for good, is that it both facilitates and indefinitely multiplies our activities. Not only does it relieve us mechanically of a crushing weight of physical and mental labor; but by the miraculous enhancing of our senses, through its powers of enlargement, penetration and exact measurement, it constantly increases the scope and clarity of our perceptions. It fulfills the dream of all living creatures by satisfying our instinctive craving for *the maximum of consciousness* with a minimum of effort! Having embarked upon so profitable a path, how can Mankind fail to pursue it?

c Thirdly and finally—the heightening of vision. One may say of the deeper vision that I have in mind that it is conveyed to our senses by the increased power of our instruments. But in a larger and more significant sense, I mean the growth of our reflective concept of the Universe, it arises irresistibly out of the mastery we have acquired of the physical springs of the world. Because of this technical control an increasing current of free energy is flowing through the human mass: energy already promoted but hitherto absorbed by the work of the hands, and also latent energy, released and in effect created by the better ordering of matter. But it becomes decidedly apparent[2] that this added energy, when it is made available to the human social organism, can only be usefully and effectively employed in one way: it must be trans-

[2] Provided we except the regressive cases of indolence (search for *well*-being as opposed to *more*-being) which momentarily crop up here and there through the excess of ease and comfort.

formed into research and creative work. The more free Man's mind is, the more does he reflect; and the more he reflects the further do his thoughts penetrate and the more intensively do they become arranged in closely related systems. That is why the great wave of modern technical progress is automatically accompanied by an ever-spreading ripple of theoretical thought and speculation. Everybody knows, without troubling to weigh the reason or importance of a fact seemingly so commonplace, that nothing is more impossible than to inhibit the growth of an idea. Applying this in its widest sense, the surest affirmation we can make about the human future is that nothing will ever restrain Man from seeking to think and essay everything to the very end.

Unification, technification, growing rationalization of the human Earth: we need to shut our eyes to the spectacle of the world we live in, it seems to me, if we are to suppose that we can ever escape from these three basic trends. But let me add at once that we must be insensitive to what, for want of a better word, I will call the "excellence" of the Universe if we are alarmed or rebellious at a prospect that it would be radically wrong to regard as a humiliating threat to our liberty. For how can we fail to discern in the simultaneous rise of Society, the Machine and Thought, this threefold tide that is bearing us upward, the essential and primordial process of Life itself—I mean, the organic in-folding of Cosmic matter upon itself, whereby ever-increasing unity, subtended by ever-heightened consciousness, is achieved by ever more complicated structural arrangements? We must not suppose, even at this early and half-passive stage of our hominization, that the partly enforced flowering of thought imposed on us by planetary pressure represents a force of enslavement of which we are the victims: we must recognize it as a force of liberation.[3]

What matters is that in the interacting development of these two basic trends upon which Mankind is continuing to build itself, technical organization and the growth of reflective consciousness, the second should acquire an ever greater predominance and degree of autonomy, conformably with the fundamental law of "vital in-folding."

And it is here, the General Tendencies being established, that the question of the Conditions of the Future arises.

2. The Conditions

IF IT IS true that, bound by the collective interaction of its liberties, the human social group cannot escape from certain irreversible laws of evolution, does this mean that, observed along its axis of "greatest complexity" (i.e., increasing liberty) the World is coiling upon itself with as much sureness as it is in other respects radiating outward and explosively expanding? In other words, because certain unalterable factors compel us to advance, with no possibility of return, in the direction of increasing hominization, must we conclude that biological evolution on Earth will easily achieve its purpose—that Thought will necessarily succeed in so shaping itself that in the end it will comprehend everything?

By no means; and for a series of reasons (or conditions to be fulfilled) which I must now set forth, from the most superficial to

---

[3] Which does not mean, alas, that the liberating process will not be accompanied by a certain amount of suffering, setbacks and even apparent wastage: the whole problem of Evil is restated (more comprehensibly, it seems to me, than in the case of a static world) in this vision of a Universe *in evolution*.

the most profound. First conditions of survival, then conditions of health, and finally, above all, conditions of synthesis.

   a   *First—Conditions of Survival.* I am not thinking particularly of the possibility of a cosmic catastrophe which might render the earth prematurely uninhabitable. The presumed duration of the whole human development (a few million years) is so trifling compared with the extent of astronomic time, even at the lowest estimate, that the chance of a variation of the solar equilibrium while anthropogenesis is in process may be ignored. Nor do I propose to dwell upon the truly negligible possibility of some rash or criminal experiment blowing up the world (there is, after all, an instinct of planetary preservation) or even of some infectious disease causing the total elimination of an animal group as far-sighted, progressive and ubiquitous as Mankind in the adult state. But on the other hand I think we must pay serious attention to warnings such as that recently uttered by Mr. Fairfield Osborn, in his book *Our Plundered Planet.*[4]

   In our hurry to advance are we not squandering our reserves to such an extent that our progress may soon be brought to a halt for lack of supplies? Where physical energy and even inorganic substances are concerned, science can foresee and indeed already possesses inexhaustible substitutes for coal, petroleum and certain metals. But foodstuffs are another matter. How long (if it ever happens at all) will it take chemical science to find ways of feeding us by the direct conversion of carbon, nitrogen and other simple elements? The population graph is rising almost vertically, while arable land in every continent is being ruined for lack of proper husbandry. We must take care: we still have feet of clay.

---

[4] Brown & Co., Boston, 1948. See also in *Harper's* (February 1948) the well-documented article by C. Lester Walker: "Too Many People."

b Secondly—Conditions of Health. I am thinking far less of hygiene and physical culture, to which sufficient thought is devoted already, than of the vital problems posed by genetics, which are willfully ignored. As I mentioned above after rising slowly until the seventeenth century, when it reached about 400 millions, the earth's population began to shoot up in an alarming fashion. It was 800 millions by the end of the eighteenth century, 1,600 millions by 1900 and over 2000 millions by 1940. At the present rate of increase, regardless of war and famine, we must expect a further 500 millions in the next twenty-five years. This demographic explosion, so closely connected with the development of a relatively unified and industrialized Earth, clearly gives rise to entirely new necessities and problems, both quantitative and qualitative. From the palaeolithic age onward, and still more after the neolithic age, Man has always lived in a state of expansion: to him to grow and to multiply have been one and the same thing. But now we suddenly see the saturation point ahead of us, and approaching at a dizzy speed. How are we to prevent this compression of Mankind on the closed surface of the planet (a thing that is good in itself, as we have seen, since it promotes social unification) from passing that critical point beyond which any increase in numbers will mean famine and suffocation? Above all, how are we to ensure that the maximum population, when it is reached, shall be composed only of elements harmonious in themselves and blended as harmoniously as possible together? Individual eugenics[5] (breeding and education designed to produce only the best individual types) and racial eugenics (the grouping or intermixing of different ethnic types being not left to chance but effected as a controlled process

---

[5] The word is used here in its general and etymological sense of "perfection in the continuance and fulfillment of the species."

in the proportions most beneficial to humanity as a whole), both, as I well know, come up against apparently insuperable difficulties, from the point of view of technical organization and from that of psychological resistance. But this does not alter the fact that the problem of building a healthy Mankind already stares us in the face and is growing more acute every day. With the help of science, and sustained by a renewed sense of our species, shall we be able to round this dangerous corner?

    c *Finally—Conditions of Synthesis, the most important of all.* What does the term mean? Cosmically speaking, as I have said, Man is collectively immersed in a "vortex" of organization which, operating above the level of the individual, gathers and lifts individuals as a whole toward the heightening of their power of reflection by means of a surplus of technical complexity. But, given the nature of the *reflexive* phenomenon, what rule must this evolutionary process observe if it is to fulfill its purpose? Essentially the following: that within the compressive arrangement which gathers them into a single complex center of vision, the human elements must group and tighten not merely without becoming distorted in the process, but with an enhancement of their "centric" qualities, i.e., their personality:[6] a delicate operation and one which, biologically, it would seem to be impossible to carry out except in an atmosphere (or temperature) of unanimity or mutual attraction. Recent totalitarian experiments seem to furnish material for a positive judgment on this last point: the individual, outwardly bound to his fellows by coercion and solely in terms of function, deteriorates and retrogresses:

---

[6] It must truly be said that this is not merely a condition of success but a positive requirement of growth. Although compelled to totalize himself (collectively) Man, at all costs, must not cease at the same time to personalize himself. This is the whole problem and drama of anthropogenesis.

he becomes mechanized. To repeat a comparison I have already used above, under these purely enforced conditions the center of consciousness cannot achieve its natural growth rising out of the technical center of social organization. Only union *through* love and *in* love (using the word "love" in its widest and most real sense of "mutual internal affinity"), because it brings individuals together, not superficially and tangentially but center to center, can physically possess the property of not merely differentiating but also personalizing the elements which comprise it. This amounts to saying that even under the irresistible compulsion of the pressures causing it to unite, Mankind will only find and shape itself if men can learn to love one another in the very act of drawing closer.

But how is this warming of hearts to be realized? In my paper on the formation of the Noosphere[7] I suggested that the very excess of external compression to which we are subjected by the relative contraction of our planet may one day cause us to breach that mysterious wall of growing repulsion which, more often than not, sets the human molecules in opposition to one another, and enter the powerful, still-unknown field of our basic affinities. In other words, attraction will one day be born of enforced nearness. I am very much less disposed to believe today that the tightening of the human mass will *of itself* suffice to warm the human heart. But I continue to believe, if anything more strongly, in the hidden existence and eventual release of forces of attraction between men which are as powerful in their own way as nuclear energy appears to be, at the other end of the spectrum of complexity. And surely it is this kind of attraction, the necessary condition of our unity, which must be linked at its root with the radiations of some ultimate Center (at once transcendent and immanent) of psychic con-

[7] See above, p. 149.

gregation: the same Center as that whose existence, opening for human endeavor a door to the Irreversible, seems indispensable (the supreme condition of the future!), for the preservation of the *will* to advance, in defiance of the shadow of death, upon an evolutionary path become reflective, conscious of the future . . .[8]

If this is true, is it not apparent that the success of Anthropogenesis, ultimately dependent upon achieving contact with the supracosmic, must, despite the rigors of its external conditioning, essentially contain an irreducible element of indeterminacy and uncertainty?

Conclusion

ALL THINGS TAKEN into account, where does the balance lie between these diverse influences, "for and against"? Faced by the biological dilemma confronting our zoological group (unite or perish) which are we to accept, which way rather than another, as the direction in which the indeterminacy essential to the human adventure is most likely to be resolved?

As I have said elsewhere, the more we study the past, noting the steady rise of Life over millions of years, and observing the ever-growing multitude of reflective elements engaged in the construction of the Noosphere; the more must we be convinced that by a sort of "infallibility of large numbers" Mankind, the present crest of the evolutionary wave, cannot fail in the course of its guided probings to find the right road and an outlet for its higher ascent. Far from being stultified by overcrowding, the cells of individual freedom, in a concerted action growing more powerful as

---

[8] See above, Chapter 13—The Human Rebound of Evolution, p. 192.

they increase in numbers, will rectify and redress themselves when they begin to move in the direction toward which they are inwardly polarized. It is reasoned calculation, not speculation, which makes me ready to lay odds on the ultimate triumph of hominization over all the vicissitudes threatening its progress.

For a Christian, provided his Christology accepts the fact that the collective consummation of earthly Mankind is not a meaningless and still less a hostile event but a precondition[9] of the final, "parousiac" establishment of the Kingdom of God—for such a Christian the eventual biological success of Man on Earth is not merely a probability but a certainty: since Christ (and in Him virtually the World) is already risen. But this certainty, born as it is of a "supernatural" act of faith, is of its nature supraphenomenal: which means, in one sense, that it leaves all the anxieties attendant upon the human condition, on their own level, still in the heart of the believer.

PARIS, JUNE 30, 1948, *PSYCHÉ*, OCTOBER 1948.

---

[9] Necessary, but not sufficient in itself.

CHAPTER 16

THE ESSENCE OF THE
DEMOCRATIC IDEA

A Biological Approach to the Problem

DEMOCRACY IS NOT an abstract concept of the
kind that can be set forth geometrically in terms of
pure ratiocination. Like so many of the notions on
which modern ideologies are based—evolution,
progress, feminism and so forth—it was originally,
and to a great extent still is, no more than the approximate expression of a profound but confused
aspiration striving to see the light and to take
shape. For this reason its elucidation calls for as
much, or more, psychology as logic. Do we not all
spend our lives in seeking to interpret ourselves by
way of actions that often appear contradictory?
How can we hope to understand ourselves without
first possessing some knowledge of our nature, history and temperament?

It may be that the growth of modern Democracy, and consequently the impulses underlying it,
will become more intelligible if, disregarding the

political and juridical aspects, we approach the problem in biological terms.

The question asked is, "What is Democracy?" Would it not be more exact and profitable slightly to modify our phraseology and to ask: "What exactly is hidden *behind* the idea of Democracy?"

1. The Present Evolutionary State of Mankind

THE PEOPLE WHO make it their business to study or order human society (ethnologists, politicians, political economists, etc.) do so in practice as though Social Man were virgin wax to be molded into any shape they choose. They do not seem to have noticed that the living substance they are manipulating is, by reason of its very formation, characterized by certain narrowly defined lines of growth; and that these, although they are sufficiently supple to permit the architects of the New Earth to make use of them, are also strong enough to disrupt any attempted arrangement that does not respect them.

This being so, of all the structural tendencies inherent in the human mass the most fundamental (indeed, the one from which all others are derived) is undoubtedly that which has led Mankind, under the twofold influence of planetary compression and psychic interpenetration, to enter upon an irresistible process of unification and organization upon itself. But to this a vital condition is attached, namely, that if it is to be viable and stable the resulting unification must not stifle but on the contrary must exalt the incommunicable uniqueness of each separate element in the unified system: something that is proved possible on a small scale by every successful team or association. In point of fact to the enlightened observer it is perfectly apparent that we could more easily prevent

the earth from turning than Mankind from progressing, laboriously but inexorably, in a twofold conjoined movement toward a
personalizing totalization. This evolutionary situation (arising out
of a very much more generalized movement of "in-folding," cosmic in its dimensions) could go unperceived while human socialization, still in its initial phase of *expansion*, was spreading over the
earth's surface. But it becomes increasingly manifest as the second
phase which we have now entered, that of socialization through
compression, takes clearer shape around us. And I believe it is this,
to the extent that it is beginning to penetrate our consciousness,
that is arousing the turmoil of so-called "democratic" aspirations
in all our hearts.

2. Biological Definition and Interpretation
of the Spirit of Democracy

LET US ASSUME that the strangely contagious modern obsession
with democratic ideas is nothing else than the feeling and liking
Man has acquired for a process which, by the collective organization of the zoological group to which he belongs, is carrying
him toward certain new states of superpersonalization—or, which
comes to the same thing, superreflection. In other words, let us
identify the spirit of Democracy with the "evolutionary sense" or
"the sense of species"—the last signifying, in the case of Man, not
merely the instinct for permanence through propagation, but also
a will to grow through the organized arrangement of the species
upon itself. We need do no more than this, it seems to me, and we
shall find that light is shed on countless points that have hitherto
been obscure, and that many disquieting antinomies have become
(at least in theory) effortlessly reconciled.

Let us apply this principle first to the legendary attributes, Liberty, Equality, Fraternity, which are indissolubly associated in our minds with the idea of any government of the people by the people; and then to the conflict, now more acute than ever, which has always divided Democracy into two factions, liberal and socialist.

a Liberty, Equality, Fraternity. It was in 1789 that this famous slogan electrified the Western world: but as events have shown, its meaning was far from clear to the minds of those it inspired. Liberty—to do *anything?* Equality—in *all* respects? Fraternity—based on *what* common bonds? . . . Even today the magical words are much more *felt* than understood. But does not their undeniable, if vague, attraction take on a clearer aspect if we consider them, as I suggest, from a biological standpoint?

Liberty: that is to say, the chance offered to every man (by removing obstacles and placing the appropriate means at his disposal) of "transhumanizing" himself by developing his potentialities to the fullest extent.

Equality: the right of every man to participate, according to his aptitudes and powers, in the common endeavor to promote, each by way of the other, the future of the individual and the species. Indeed, is it not this need and legitimate demand to *participate* in the Human Affair (the need felt by every man to live coextensively with Mankind) which, deeper than any desire for material gain, is today agitating those classes and races that have hitherto been left "out of the game"?

Fraternity: as between man and man, in the sense of an organic interrelation based not merely on our more or less accidental coexistence on the surface of the earth, or even on our common origin, but on the fact that we represent, all of us together, the *front line*, the crest of an evolutionary wave still in full flood.

Liberty, Equality, Fraternity—no longer indeterminate, amorphous and inert, but directed, guided, dynamized by the growth of a fundamental impulse which underlies and sustains them.

Does not everything truly become more clear in the light of this guiding principle?

b *Liberal Democracy and Directed Democracy*. The UNESCO questionnaire refers in passing to the disparity, deplored by de Tocqueville, between "democracy" and "socialism." Broadly speaking, the avowed object of the inquiry is an attempt to resolve, at least theoretically, the present tensions in this field between East and West.

But does not the strange and persistent cleavage, so invariably manifest within so-called democratic movements in the opposed concepts of liberalism and *dirigisme* (or individualism and totalitarianism) explain itself when we realize that, although they may look like contradictory social ideals, they are in fact natural components (personalization and totalization) whose interaction biologically determines the essence and progress of anthropogenesis? On the one hand we have a system centered on the individual, and on the other a system centered on the group. Sometimes the first of the two vectors, sometimes the second, breaks away and so dominates the other as to appear determined to engulf everything. A shift to the right is followed by a shift to the left. But there is really no fundamental contradiction in this. It is simply a matter of disconnection and disharmony which may even (why not?) be an inevitable and necessary alternation. Biologically, let me repeat, there can be no true Democracy without the balanced combination of these two complementary factors, which in their pure state are expressed, one by individualist and the other by authoritarian regimes.

But in practical terms how precisely are we to proceed eventually in order to bring them into harmony?

3. *The Technique of Democracies*

VERY PROPERLY, A large number of UNESCO's questions are concerned with the study and criticism of the existing forms and methods of Democracy. Since this is a sphere in which I have no competence I shall confine myself to the three following remarks, all from a biological standpoint:

a In the first place, and in the light of what I have said, there are two *general conditions* which must at all costs be observed in the planning of democratic institutions. The first of these is that the individual must be allowed the widest possible liberty of choice within which to develop his personal qualities (the one theoretical restriction being that his choice should be exercised in the direction of heightened powers of reflection and consciousness). The second, off-setting the first, is that everything must be done to promote and foster the currents of convergence (collective organizations) within which alone, by the laws of anthropogenesis, individual action can achieve its fulfillment and full consistence. In short, what is needed is a judicious mixture of *laissez-faire* and firmness. The problem is one of moderation, tact and "art" for which no hard-and-fast rules can be laid down, but which, in each particular case, every nation is perfectly capable of solving in its own way—provided its instinct of progress and "superhumanization" is sufficiently developed.

b Secondly, it is only by way of countless *experiments and gropings* that the Democratic ideal (like Life itself) can hope to achieve its own formulation and, still more, can materialize. Despite the compressive and unifying conditions to which we are subject, Mankind is still made up of terribly heterogeneous parts, unequally matured, whose democratization can be effected only with the use of imagination and suppleness, and in conformity with the varying circumstances in each portion of the World.

c Finally, it is upon the maintenance and growth in human consciousness of what I have called the "sense of the Species" that the realization of a truly democratic world society ultimately depends. Only a powerful polarization of human wills, after each fragment of humanity has been led to the discovery of his own particular form of freedom, can ensure the convergence and unified working of this plurality in a single, coherent planetary system. Above all, only this polarization, through the unity thus constituted, can create the atmosphere of noncoercion—*unanimity*—which is, when all is said, the rare essence of Democracy.

UNPUBLISHED. PARIS, FEBRUARY 2, 1949.

IN RESPONSE TO A QUESTIONNAIRE FROM UNESCO.

CHAPTER 17

DOES MANKIND MOVE BIOLOGICALLY UPON ITSELF?

Galileo's Question Restated

IN THE PAST three years I have twice sought in these pages[1] to depict and interpret from a purely scientific standpoint what I have called "the organic concretion upon itself of Mankind" and the corresponding "rebound," in terms of biological evolution, that seems to result from this.

I beg leave to return to the subject, this time not with the calmness of a theorist adding to his argument but with a greater and fiercer vehemence, the better to stress the vital importance and urgency, both for our thought and our action, of the problem presented by the explosion of human Totalization which we already see in full spate around us.

Describing the formation of a thinking enve-

[1] *Revue des Questions Scientifiques,* January 1947 and April, 1948. In this volume, "The Formation of the Noosphere," p. 149: and "The Human Rebound of Evolution and its Consequences," p. 192.

lope, a Noosphere, now being shaped round our planet, I wrote that this was a "defensible hypothesis." But did I speak strongly enough? Was there not a certain perfidy in those soothing and cautious words, even a hint of cowardice? I wrote of "plausible views," as though all this were no more than a game of academic speculation inviting no intellectual commitment, to be taken up or dropped at our leisure, with all conclusions deferred. But does this attitude of unconcerned detachment really meet the situation of the individual man today, who finds himself confronted by the expansion and over-flowing of human collectivization? Surely the truth, for those of us who seek to understand the portents we see multiplying around us, is that we must face the fact that in *no sphere*, whether politico-economic or social, artistic or mystical, can anything stable or en-during be built on Earth until we have found a positive answer to the following question:

What degree of reality and what ontological significance are we to attribute to this strange shift of the current, as a result of which modern man, scarcely entered into what he supposed to be the haven of his individual rights, finds himself suddenly drawn into a great unitary whirlpool where it seems that his most hard-won attributes, those of his incommunicable, personal being, are in danger of being destroyed? Is it Life that we see on the horizon, concealed behind the rise of the masses, or is it Death?

In recent months my observation and personal knowledge have made me increasingly conscious of a paramount and imme-diate necessity: the necessity for our generation of adopting certain firm values regarding the course of the world, of taking a major decision upon which the future of human history will depend.

It is this dramatic situation whose urgency I wish to stress in the form of three essentially related propositions of which each of us carries the substance in his heart, without choosing, or without daring, to acknowledge them and accept their inexorable pressure and their natural logic.

1. The First Point (beyond dispute): The Material Fact of the In-folding of Mankind upon Itself

AS A BASIS for what I have to say (indeed, as a basis for any attempt to understand what is happening in the world) a plain fact must be plainly stated which perhaps my previous writings have failed to detach sufficiently from certain accessory hypotheses—although, rid of encumbrances, it is as evident as the rising sun. I mean the essentially modern fact of the "social-scientific agglomeration" of Mankind upon itself.

Let us first be sure that we understand the meaning of the word "modern." We can all see, at least in retrospect, that since Man first appeared on earth he has not only spread over the continents in an amorphous flood but has constituted himself a group of exceptional solidarity, not merely ubiquitous but "totalitarian" in tendency. But during tens of thousands of years the weaving and extension of the thinking network over the surface of the globe proceeded so slowly and sporadically that until quite recently not even the most acute observers, although they recognized the biological singularity of our nature, seem to have suspected that, zoologically speaking, Humanity might be wholly unique in its destiny and structural potentialities. To the old naturalists the human "species" had virtually achieved its vital equilibrium, and they found nothing (except of course a more highly developed mind and sensibility) to distinguish it from other animal families where the probable curve of its development was concerned.

Such was the comfortable vista which, less than a century ago, began abruptly to change beneath our gaze, something in the fashion of those organic tissues in the living body which, after long remaining harmless and dormant, their cells apparently indistinguishable from those of the surrounding tissue, suddenly burst into dangerous growth. Or we may liken the change to the fall of an avalanche in the mountains—a sudden calamity, long and noiselessly prepared—or, better still, to the sudden birth of a cyclone in tranquil, overheated air, or of a whirlpool in the smooth waters of a river. Is not this precisely the kind of spectacle to which we are now awakening (still without really believing in it); the spectacle of Mankind, suddenly shaken out of a deceptive inertia, being swept along ever more rapidly, by the current of its own proliferation and contraction, into the diminishing circles of a sort of maelstrom coiling it irresistibly upon itself?

Let us try and get some idea of the speed (the rising curve, if you prefer) of this process of in-folding over the period of a single generation. Looking back to the turn of the century we see limited wars, clearly marked frontiers, large blank spaces on the map and distant, exotic lands, to visit which was still like entering another world. Today we have a planet girdled by radio in the fraction of a second and by the airplane in a few hours. We see races and cultures jostling one another, and a soaring world population amid which we are all beginning to fight for elbowroom. We see a world, stretched almost to breaking-point between two ideological poles, where it is impossible for the smallest peasant in the remotest countryside to live without continually being in some way affected or worried by what is going on in New York or Moscow or China . . .

To steady ourselves in face of this progressive invasion of our private lives, and keep our footing in the rising tide, we tend to deny the reality of what is happening. Or we tell ourselves that this intermingling of all men over all the earth can portend no more than a passing phase of political readjustment such as has already occurred many times in history: not a definitive and lasting phenomenon, but a momentary complexity of circumstances, brought about by chance at the present time, and later to be resolved by some other chance.

But it must surely be clear that in this we are simply deceiving ourselves. Is it conceivable that the human world will relax its grip, loosen the coils which it has woven round all our separate lives? How can we dare to suppose it?

No matter where we look, there is no indication that the grip is tending to relax. I do not deny that there are revolts against it, on the part of individuals and even of nations; but these spasmodic movements of protest are painfully crushed by the tightening of the vise almost as soon as they appear.

Nor is this all. As the phenomenon of human in-folding takes shape, so does its implacable and henceforth unchangeable mechanism become apparent. In origin this is a material force, that of terrestrial compression acting on a rapidly growing population composed of elements whose field of action grows even more rapidly. The Earth is visibly contracting, and the hundreds of millions of people on its surface, reacting to the pressure of this contraction, are compelled not only to make technical arrangements among themselves, but to accept and use the inexhaustible spiritual and intellectual links born of the revolutionary power of Reflection. How can we hope to resist, how dream of escaping, from the play of these two cosmic coils (spatial and mental) that close in upon us in a movement conjugated upon ourselves? The human molecules are tightly packed together, and the more this is the case the more impossible it becomes for them, owing to their nature and structure, not to merge both physically and in spirit. Rising social standards, the rise of the Machine, the growth of knowledge . . . We are naively disposed to speak, to marvel, or even grow indignant, as though these separate events and their conjunction were something accidental or surprising. But how can we fail to see that we are simply dealing with three aspects of a single perfectly regulated process on a *planetary scale?*

It is amazing how often, in casual reading or conversation, one encounters either a total inability to see and understand, or a deliberate refusal to accept, the plainest evidence of this material fact of the inevitable drawing-together of Mankind. Let us consign past thinking to the past. Yesterday, perhaps, it was still possible for us to wonder whether Mankind as an ethnic and cultural whole could be said to constitute a finally stabilized group: today, overtaken by the rush of events, there is no longer room for any uncertainty in the matter. For whatever deeper reasons, still calling for discovery and debate, the human mass, which at one time seemed immobile or immobilized, is again on the move. The wheels are in motion and the speed is visibly increasing, rendering perfectly abortive any attempt at resistance on our part, whether physical or mental, to the tide that is sweeping us along. For however long it may endure, the human world will henceforth only be able to continue to exist by organizing itself ever more tightly upon itself. We may delude ourselves with the notion that we are simply weathering a storm. The truth is that we are undergoing a radical change of climate.

We must accept it once and for all. Human Problem No. 1 is no longer that of deciding whether we can escape the sociophysical infolding of the human race upon itself, since this is irrevocably imposed on us by the physiochemical structure of the Earth. All that matters, the only meaningful question, is to know whither this process of totalization is leading us—toward what summit, or what abyss?

2. The Second Point (which now emerges): The Biological Value of Human Socialization

WHAT RESULTS FROM the foregoing is that, confronted by this technico-social embrace of the human mass, modern man, in so far as he has any clear idea of what is happening, tends to take fright as though at an impending disaster. Having scarcely emerged, after millennia of painful differentiation, from what the ethnologists call the state of primitive co-consciousness, are we now, through the very excess of our civilization, to sink back into a state of even greater obscurity?

It must be said that appearances (or excuses) are not lacking to warrant this pessimistic view. At close quarters and on the individual level we see the ugliness, vulgarity and servitude with which the growth of industrialism has undeniably sullied the poetry of primeval pastures. At a higher level we see the somber threat, still increasing despite the surgical operation of the second World War which was supposed to abate it, of so-called political totalitarianism. And on what is, in a sense, a higher level still we have the disquieting example of such animal groups as termites, ants and bees, our ancestors in the Tree of Life, which, afflicted by an evil of which we seem to perceive the symptoms in ourselves, have lapsed into a state of social enslavement—the very fate toward which an implacable destiny seems to be impelling us. Evidence such as this, if it is insufficiently studied, must certainly cause us dismay. Does it not suggest that this is a general law of life; that the living creature, compelled for its own survival to attach itself materially and spiritually to others of its kind, and to an increasing extent as it progresses autonomously and in individual freedom, is automatically prevented by Nature from rising above a given level of emancipation and consciousness? And may it not be that we are now thrusting against this barrier the surface-limit of the "I"?

If that is so, then the whirlpool in which we are seized must end by dehumanizing us. It means that we are lapsing into a sort of senescence. This is one way of explaining the irresistible movement of concentration that has us in its grip: the quick answer, the simplest and most morbidly fascinating to sensibilities as shaken and bruised as our own at the present time.[2]

But (let it be cried from the rooftops!) there is another way. To that tired and outworn approach to the problem of the Socialization of the Earth we may oppose an entirely different, very much more constructive interpretation, solidly and scientifically based on a new vision of Life and the World, of the grand phenomenon of human Totalization.

Let me briefly recapitulate the theory in its broad outline, looking first at Life as Science is now beginning to rediscover and reassess it.

Because plants and animals are excessively fragile in their structure, and because until now their existence has only been detected, indeed is only conceivable, in severely limited zones of Time and Space, we have been accustomed to regard them as an anomaly and an exception, almost a small, separate world within the great universe. But much has happened to change this view. New lights have been vouchsafed us—the reality, capable of definition, of a Cosmogenesis; the discovery of the genesis of atoms, of the increasingly "molecular" aspect of living organisms pursued to the infinitesimal, and of the persistence of this "molecular" characteristic in the mechanisms of heredity and evolution rising to the highest organic types; the existence of a center of indeterminacy at the very heart of every element of Matter . . . The cumulative effect of these revelations has been to open our eyes to a very different and quite otherwise alluring possibility. This is that living beings, far from constituting a singular and inexplicable oddity in the world, are on the contrary the final outcome of an entirely generalized

[2] Three reasons, among others, which explain the favor it enjoys in contemporary literature and in the conservative and existentialist press. It is so easy to write and get oneself read if one sets up as a prophet of disaster. "Frighten me. . . ."

physicochemical process in virtue of which the stuff of the cosmos, by virtue of the very position it occupies and its structure, is not only in a state of spatial expansion (as in these days is generally accepted) but that, even more significantly, it presents itself to our experience as actuated by a movement of qualitative in-folding (or arrangement, if you prefer) upon itself; and this "in-folding arrangement" moves in the direction, not of any homogeneous repetition, but of a formidable growth of complexity, increasing with the passage of time and resulting in proteins, cells and living matter of every kind. A certain Laplacian cosmology having accustomed our minds to the idea that the phenomena of the dissipation of energy and the way of "greatest probability" are the only ones physically possible, we instinctively recoil from the thought of a partial "lapse" of the Universe into Complexity. Yet, just as the shift to red of the galactic spectra compels us to accept the centrifugal flight of the sidereal layers—so, and still more certainly, must the distribution and history of the small, the big, and, finally, the very large particles (what is each of us, in effect, but an immense molecule?) point to a continuous global drift of the "stuff of things" toward ever more advanced and bewilderingly elaborate types of construction that are closed and centered on themselves?[3]

Viewed in this way Life, far from being an aberration on the part of Nature, becomes within the field of our experience nothing less than the most advanced form of one of the most fundamental currents of the Universe, in process of taking shape around us. Which is to conclude that, everywhere exerting its pressure, it tends with a "cosmic" tenacity and intensity to make continuous progress wherever it has gained a foothold, always moving in the same direction and reaching out as far as possible.

[3] Thus matter has a "ballast" of, not so much geometry (as Bergson maintained), as complexity. The fact may be hard to explain, but it is there before our eyes.

Having postulated so much, let us return to the significance and value of human totalization.

What characterizes this process in our eyes, as we have said, is the network of social-technical bonds from which our personal freedom suffers at the first contact, and in which, at first sight, we see nothing but diminution and mechanization.

But why do we, or more precisely *how can we*, fail to detect in this same process, and to the exact extent that it constitutes an organized whole—how fail to perceive, beneath what looks at first sight like a purely blind mechanical operation, the same process of "complexification" which, if our new concept is valid, constitutes the essential proceeding of Life? What discloses and measures, both as a reagent and a parameter, the *eu-complex* or organic arrangements of matter, as opposed to the purely fortuitous or mechanical (*pseudo-complex*) groupings that take place between atoms and molecules, is the appearance and growth in the former of psychic properties. In physicochemical terms a very high degree of complexity begets consciousness: few laws of Nature are so sure, so consistent (or so little exploited) as this. And whatever may be said about the "gregariousness of crowds" and the "brutalization of the masses"[4] the fact remains beyond dispute that, although human socialization may not yet have shown itself to be particularly productive of virtue,[5] it is this process that has unloosed the formidable scientific impulse that is causing us to revise our conception of the Universe from top to bottom.[6]

---

[4] Effects, not of complexity, but of unorganized large numbers.

[5] We must wait and see.

[6] Without, of course, denying the great fact of the growth of science, the "pessimists" tend to see in it nothing but an accidental (and dangerous) by-product of social mechanization—something utterly without spiritual value: "epimind," one might call it, released by an epiphenomenon.

Human Totalization develops mind; it goes hand-in-hand with "psychogenesis": THEREFORE it is of a biological nature, order and dimension.

We need no further evidence, in my view, to prove scientifically that the social in-folding which we are undergoing is nothing other than the direct and logical extension, over our heads, of the process of cosmic in-folding which gave birth to the first cell and the first thought on earth. Supercomplexification and superinteriorization, within the zone that I have called the Noosphere, of the stuff of the Universe: not only of *men*, but of the *Man* who is to be born tomorrow! Everything falls into place around us, amid the so-called human chaos, if we view it in this light. The World goes on its way—and that is all.

We rebel at yielding to the excess of vitalization that is bearing us along because we fear that in doing so we shall lose the precious fragment, "me", which we have acquired. But how can we fail to see that, of itself and properly controlled,[7] and provided it acts not simply on what is "mechanizable" (instinct) but on what can be rendered "unanimizable" (reflective psychism) totalization by its nature does not merely differentiate but *personalizes* what it unites?

All things considered it seems, then, that we have two opposed evaluations of this process of the in-folding of Humanity upon itself; a process which is clearly apparent and which nothing can prevent.

a That the planetary collectivization which confronts us is a crude phenomenon of mechanization or senescence which will end by dehumanizing us;

b That on the contrary it is a mark and an effect of biological superarrangement destined to ultrapersonalize us.

[7] Within a field of affective attraction sufficiently intense to influence the human mass as a whole and at the same time.

I maintain that of these two judgments it is the second which is far more likely to fit the reality. That its validity has not yet been finally demonstrated or verified may be admitted. Nevertheless it is solid enough for us to feel, considering the moment of history in which we live, that the time has come when we may, indeed *must*, resolutely stake our future on it.

This brings me to the heart of what I am trying to say in these pages.

The Hour of Choice

IN EVERY SPHERE, physical no less than intellectual and moral, and whether it be a question of flowing water, a traveler on a journey, or a thinker or mystic engaged in the pursuit of truth, there inevitably comes a point in time and place when the necessity presents itself, to mechanical forces, or to our freedom of choice, of deciding once and for all which of two paths is the one to take. The enforced, irrevocable choice at a parting of the ways that will never occur again: which of us has not encountered that agonizing dilemma? But how many of us realize that it is precisely the situation in which social man finds himself, *here and now*, in face of the rising tide of socialization?

Borne on a current of Totalization that is taking shape and gathering speed around us, we cannot, as I have said, either stop or turn back. Indeed, how can we even contemplate escaping from a tide that is not only planetary but cosmic in its dimensions?

As I have also shown, two attitudes are theoretically possible in this situation, two forms of "existentialism." We can reject and resist the tide, seeking by every means to slow it down and even to escape individually (at the risk of perishing in stoical isolation) from what looks like a rush to the abyss; or we can yield to it and actively contribute to what we accept as a liberating and life-giving movement.

It remains for me to demonstrate the urgency of the problem; that is to say, to fulfill my purpose by showing that we have truly reached the parting of the ways, the point where the waters divide; and also to show that in this momentous hour we cannot continue physically to exist (to act) without deciding here and now which of the two attitudes we shall adopt: that of defiance or that of faith in the unification of mankind.

The urgency is due first and foremost to the state of deep-seated irresolution created by our seeming lack of choice in face of the immense problems which Mankind must solve without delay if it is to survive. We debate endlessly about Peace, Democracy, the Rights of Man, the conditions of racial and individual eugenics, the value and morality of scientific research pushed to the uttermost limit, and the true nature of the Kingdom of God; but here again, how can we fail to see that each of these inescapable questions has *two aspects*, and therefore *two answers*, according to whether we regard the human species as culminating in the individual or as pursuing a collective course toward higher levels of complexity and consciousness? Let such organizations as the U.N. and UNESCO continue to multiply and flourish; I for one shall always rejoice unreservedly in their existence. But we must realize that we shall be forever building on shifting sand so long as bodies of this kind are not agreed on the basic values and purpose underlying their projects and decisions—that is to say, on their attitude toward human totalization. What good does it do to discuss the ripples on the surface while the undertow is still uncontrolled?

Without realizing it—and this is at the root of our present political and ideological stagnation—we are still desperately clinging to the old concept, now become unliveable, of a world in a state of human immobility, as though this idea were not visibly crumbling. In doing so we run a twofold risk, not only of continuing in a state of inefficiency and chaos but, which is far worse, of missing what may be the only chance offered to the earth of achieving its maturity.

For herein lies the tragic nature of a dilemma presenting itself, as it does to Man, in purely "reflective" terms. We cannot wait passively upon the statistical play of events to decide for us which road the world is to take tomorrow. We must positively and ardently take a hand in the game ourselves. If it is true, as I suggest, that salvation lies in the direction of an Earth organically in-folded upon itself, it is then surely evident that, through a reciprocal mechanism of action and reaction, the vision and prevision of this ultimate end, this outcome of History and of Life, may be made to play an essential part in the building of the future, if only by creating the atmosphere, the psychic *field of attraction*, without which it will be impossible for Humanity to continue to converge upon itself. Again, if it be true that Evolution is rebounding on itself through the fact of human totalization, it must, becoming conscious, fasten passionately upon itself: which is to say that Man to progress further, will need to be sustained by a powerful collective faith. According to whether we believe in it or reject it, the totalizing process, from which there is no escape, will either infuse new life into us or destroy us—that is the fact. It is precisely in order to discover and bring into the open this saving and transforming Faith that we must, at this crucial instant, take a positive stand on the spiritualizing and humanizing value of social totalization, and thus reaffirm our sense of the Species on a new plane. We must do so now. Life will not wait for us, and our state is insecure. Who can say whether tomorrow will not be too late?

What is, in fact, happening in the world today is as though, four hundred years later and at a higher turn of the spiral, we found ourselves back in the intellectual position of the contemporaries of Galileo. For the men of the sixteenth century it was the idea of a Universe in motion (in the oversimplified but still perfectly recognizable form of an "absolute" rotation of the earth round the sun) which, as it dawned on their minds, affected them even more deeply than the ending of the geocentric concept. Under the influence of this initial shock, as we can now see, the whole sidereal Cosmos, as expressed in terms of the physics, philosophy and theology of those days, began to give way to a Cosmogenesis: a transformation that was no doubt less heavily loaded with practical consequences, or less directly a stimulant of action, than the one we are now undergoing, which is a passage from the concept of a static and dispersed humanity to one of humanity biologically impelled toward the mysterious destiny of a global anthropogenesis. It was a transformation of the same order of magnitude, and psychologically of the same kind: that is to say, an acquiescence, at once free and enforced, in the necessity of adopting a new point of view to the extent that this resulted from a general and irresistible maturing of the human consciousness. Following the moment when a few men began to see the world through the eyes of Copernicus all men came to see it in the same fashion. A first flash of illumination, intuitively accepted despite the risk of error; and as the intuition was increasingly confirmed by observation and experiment, it came to be embodied in the inherited core of human consciousness. In the sixteenth century—as had already happened at long intervals in the course of history, *and as is again happening today*—men found themselves suddenly "up against a blank wall," in the sense that they felt instinctively that they could not continue to be men without adopting a positive position toward a given interpretation of the phenomenal framework enclosing them. Accordingly they made their choice. And, as we look back we see that Life, reaching a major fork in the road, and acting in men and through men, once again took the right way.

May it happen—I have no doubt that it will, because I am profoundly convinced of the essential bond of complicity uniting Life, Truth and Freedom—may it happen that our descendants four centuries hence, being faced by some new parting of the ways that we cannot yet foresee, will look back and say: "In the twentieth century they saw clearly. Let us seek to follow their example!"

Once again, as in the days of Galileo and the problem of the movements of the earth, we have to achieve unanimity, this time regarding the value, whether materially constructive or vitalizing, of the phenomenon of socialization. But if I am not mistaken the balance is already swinging in favor of the organic nature (and the resultant biological effects, which we cannot yet foresee) of human "planetization." The more we allow ourselves to believe in this possible superorganization of the world, the more shall we find reason to believe in it, and the more numerous will become the believers. It seems that already a collective intuition in that direction, which nothing will be able to arrest, is on the move.[8]

So that it requires no great gift of prophecy to affirm that, within two or three generations, the notion of the psychic infolding of the earth upon itself, in the bosom of some new "space of complexity," will be as generally accepted and utilized by our successors as the idea of the earth's mechanical movement round the sun, in the bosom of the firmament, is accepted by ourselves.

SAINT-GERMAIN-EN-LAYE, MAY 4, 1949.

REVUE DES QUESTIONS SCIENTIFIQUES, OCTOBER 20, 1949.

---

[8] We must bear in mind that although the horizon in the direction of human totalization remains politically, economically and psychologically obscure, this is of little importance. The immediate question is not one of knowing precisely whither the current is taking us and how we shall shoot the rapids, but simply of deciding, having reached the fork, whether we are following the main course of the stream—that is to say, are not detaching ourselves from the World in motion.

CHAPTER 18

THE HEART OF
THE PROBLEM

Some say, "Let us wait patiently until the Christ returns."
Others say, "Let us rather finish building the Earth."
Still others think, "To speed the Parousia,
let us complete the making of Man on Earth."

Introduction

AMONG THE MOST disquieting aspects of the
modern world is its general and growing state of
dissatisfaction in religious matters. Except in a humanitarian form, which we shall discuss later, there
is no present sign anywhere of Faith *that is expanding*: there are only here and there, creeds that at the
best are holding their own, where they are not positively retrogressing. This is not because the world
is growing colder: never has it generated more psychic warmth! Nor is it because Christianity has lost
anything of its absolute power to attract: on the
contrary, everything I am about to say goes to
prove its extraordinary power of adaptability and
mastery. But the fact remains that for some obscure
reason something has gone wrong between Man
and God *as in these days He is represented to Man.* Man

would seem to have no clear picture of the God he longs to worship. Hence (despite certain positive indications of rebirth which are, however, still largely obscured) the persistent impression one gains from everything taking place around us is of an irresistible growth of atheism—or more exactly, a mounting and irresistible de-Christianization.

For the use of those better placed than I, whose direct or indirect task it is to lead the Church, I wish to show candidly in this paper *where*, in my view, the root of the trouble lies, and *how*, by means of a simple readjustment at this particular, clearly localized point, we may hope to procure a rapid and complete rebound in the religious and Christian evolution of Mankind.

I say "candidly." It would be presumptuous on my part to deliver a lecture, and criticism would be out of place. What I have to offer is simply the testimony of my own life, a testimony which I have the less right to suppress since I am one of the few beings who can offer it. For more than fifty years it has been my lot (and my good fortune) to live in close and intimate professional contact, in Europe, Asia and America, with what was and still is most humanly valuable, significant and influential—"seminal" one might say—among the people of many countries. It is natural that, by reason of the unusual and exceptional contacts which have enabled me, a Jesuit (reared, that is to say, in the bosom of the Church) to penetrate and move freely in active spheres of thought and free research, I should have been very forcibly struck by things scarcely apparent to those who have lived only in one or other of the two opposed worlds, so that I feel compelled to cry them aloud.

It is this cry, and this alone, which I wish to make heard here—the cry of one who thinks he sees.

1. *A Major Event in Human Consciousness: The Emergence of the "Ultra-Human"*

ANY EFFORT TO understand what is now taking place in human consciousness must of necessity proceed from the fundamental change of view which since the sixteenth century has been steadily exploding and rendering fluid what had seemed to be the ultimate stability—our concept of the world itself. To our clearer vision the universe is no longer an Order but a Process. The cosmos has become a Cosmogenesis. And it may be said without exaggeration that, directly or indirectly, all the intellectual crises through which civilization has passed in the last four centuries arise out of the successive stages whereby a static *Weltanschauung* has been and is being transformed, in our minds and hearts, into a *Weltanschauung* of movement.

In the early stage, that of Galileo, it may have seemed that the stars alone were affected. But soon the Darwinian stage showed that the cosmic process extends from sidereal space to life on earth; with the inevitable result that, in the present phase, Man finds himself overtaken and borne on the whirlwind which his own science has discovered and, as it were, unloosed. From the time of the Renaissance, in other words, the cosmos has looked increasingly like a cosmogenesis; and now we find that Man in his turn is tending to be identified with an anthropogenesis. This is a major event which must lead, as we shall see, to the profound modification of the whole structure not only of our Thought but of our Beliefs.

Many biologists, and not the least eminent among them (all being convinced, moreover, that Man, like everything else, emerged by evolutionary means, i.e., was *born* in Nature) undoubtedly still believe that the human species, having attained the level of *Homo sapiens*, has reached an upper organic limit beyond which it cannot develop, so that anthropogenesis is only of retrospective interest in the past. But I am convinced that, in opposition to this wholly illogical and arbitrary idea of arrested hominization, a new concept is arising, out of the growing accumulation of analogies and facts, which must eventually replace it. This is that, under the combined influence of two irresistible forces of planetary dimensions (the geographical curve of the Earth, by which we are physically compressed, and the psychic curve of Thought, which draws us closer together), the power of reflection of the human mass, which means its degree of *humanization*, far from having come to a stop, is entering a critical period of intensification and renewed growth.

What we see taking place in the world today is not merely the multiplication of *men* but the continued shaping of *Man*. Man, that is to say, is not yet zoologically mature. Psychologically he has not spoken his last word. In one form or another something ultrahuman is being born which, through the direct or indirect effect of socialization, cannot fail to make its appearance in the near future: a future that is not simply the unfolding of Time, but which is being constructed in advance of us . . . Here is a vision which Man, we may be sure, having first glimpsed it in our day, will never lose sight of.

This having been postulated, do those in high places realize the revolutionary power of so novel a concept (it would be better to use the word "doctrine") in its effect on religious Faith? For the spiritually minded, whether in the East or the West, one point has hitherto not been in doubt: that Man could only attain to a fuller life by rising "vertically" above the material zones of the world. Now we see the possibility of an entirely different line of progress. The long dreamed-of Higher Life, the Union, the consummation that has hitherto been sought *Above*, in the direction of some kind of transcendent (see OY, diagram p. 269): should we not rather look for it *Ahead*, in the prolongation of the inherent forces of evolution (see OX, diagram p. 269)?

Above or ahead—or both? . . .

This is the question that must be forced upon every human conscience by our increasing awareness of the tide of anthropogenesis on which we are borne. It is, I am convinced, the vital question, and the fact that we have thus far left it unconfronted is the root cause of all our religious troubles; whereas an answer to it, which is perfectly possible, would mark a decisive advance on the part of Mankind toward God. That is the heart of the problem.

2. At the Source of the Modern Religious Crisis: A Conflict of Faith Between the Above and the Ahead

ARISING OUT OF what I have said, the diagram at the end of this chapter represents the state of tension which has come to exist more or less consciously in every human heart as a result of the seeming conflict between the modern forward impulse (ox), induced in us all by the newly born force of transhominization, and the traditional upward impulse of religious worship (OY). To render the problem more concrete it is stated in its most final and recognizable terms, the coordinate OY simply representing the Christian impulse and OX the Communist or Marxist impulse[1] as these are commonly manifest in the present-day world. The question is, how does the situation look, here and now, as between these opposed forces?

[1] An unfavorable simplification where OX is concerned, inasmuch as Marxism and Communism (the latter a thoroughly bad, ill-chosen word, it may be said in passing) are clearly no more than an embryonic form, even a caricature, of a neohumanism that is still scarcely born.

We are bound to answer that it looks like one of conflict that may even be irreconcilable. The line OY, faith in God, soars upward, indifferent to any thought of an ultra-evolution of the human species, while the line OX, faith in the World, formally denies (at least in words) the existence of any transcendent God. Could there be a greater gulf, or one more impossible to bridge?

Such is the appearance: but let me say quickly that it cannot be true, not finally true, unless we accept the absurd position that the human soul is so badly devised that it contradicts within itself its own profoundest aspirations. Let us look more closely at OX and OY and see how these two vectors or currents appear and are at present behaving in their opposed state. Is it not apparent that both suffer acutely from their antagonism, and therefore that there must be some way of overcoming their mutual isolation?

Where OX is concerned the social experiment now in progress abundantly demonstrates how impossible it is for a *purely immanent* current of hominization to live wholly, in a closed circuit, upon itself. With no outlet ahead offering a way of escape from total death, no supreme center of personalization to radiate love among the human cells, it is a frozen world that in the end must disintegrate entirely in a Universe without heart or ultimate purpose. However powerful its impetus in the early stages of the course of biological evolution into which it has thrust itself, the Marxist anthropogenesis, because it rules out the existence of an irreversible Center at its consummation, can neither justify nor sustain its momentum to the end.

Worldly faith, in short, is not enough in itself to move the earth forward: but can we be sure that the Christian faith in its ancient interpretation is still sufficient of itself to carry the world upward?

By definition and principle it is the specific function of the Church to Christianize all that is human in Man. But what is likely to happen (indeed, is happening already) if at the very moment when an added component begins to arise in the *anima naturaliter christiana*, and one so compelling as the awareness of a terrestrial "ultra-human," ecclesiastical authority ignores, disdains and even condemns this new aspiration without seeking to understand it? This, and simply this, that Christianity will lose, to the extent that it fails to embrace as it should *everything that is human on earth*, the keen edge of its vitality and its full power to attract. Being for the time *incompletely humanized* it will no longer fully satisfy even its own disciples. It will be less able to win over the unconverted or to resist its adversaries. We wonder why there is so much unease in the hearts of religious and of priests, why so few deep conversions are effected in China despite the flood of missionaries, why the Christian Church, with all its superiority of benevolence and devotion, yet makes so little appeal to the working masses. My answer is simply this, that it is because at present our magnificent Christian charity lacks what it needs to make it decisively effective, the sensitizing ingredient of *Human* faith and hope without which, in reason and in fact, no religion can henceforth appear to Man other than colorless, cold and inassimilable.

OY and OX, the Upward and the Forward: two religious forces, let me repeat, now met together in the heart of every man; forces which, as we have seen, weaken and wither away in separation . . . Therefore, as it remains for me to show, forces which await one thing alone—not that we should choose between them, but that we should find the means of combining them.

3. The Rebound of the Christian Faith: Upward by Way of Forward

IT IS GENERALLY agreed that the drama of the present religious conflict lies in the apparent irreconcilability of two opposed kinds of faith—Christian faith, which disdains the primacy of the ultra-human and the Earth, and "natural" faith, which is founded upon it. But is it certain that these two forces, neither of which, as we have seen, can achieve its full development without the other, are really so mutually exclusive (the one so antiprogressive and the other so wholly atheist) as we assume? Is this so if we look to the very heart of the matter? Only a little reflection and psychological insight is required to see that it is not.

On the one hand, neo-human faith in the World, to the extent that it is truly a Faith (that is to say, entailing sacrifice and the final abandonment of self for something greater) necessarily implies an element of worship, the acceptance of something "divine."[2] Every conversation I have ever had with communist intellectuals has left me with a decided impression that Marxist atheism is not absolute, but that it simply rejects an "extrinsicist" form of God, a *deus ex machina* whose existence can only undermine the dignity of the Universe and weaken the springs of human endeavor—a "pseudo-God," in short, whom no one in these days any longer wants, least of all the Christians.

And on the other hand Christian faith (I stress the word Christian, as opposed to those "oriental" faiths for which spiritual ascension often expressly signifies the negation or condemnation of the Phenomenon), by the very fact that it is rooted in the idea of Incarnation, has always based a large part of its tenets on the tangible values of the World and of Matter. A too humble and subordinate part, it may seem to us now (but was not this inevitable in the days when Man, not having become aware of the genesis of the Universe in progress, could not apprehend the spiritual possi-

---

[2] As in the case of biological evolutionary theory which also bore a materialist and atheist aspect when it appeared a century ago, but of which the spiritual content is now becoming apparent.

bilities still buried in the entrails of the Earth?) yet still a part so intimately linked with the essence of Christian dogma that, like a living bud, it needed only a sign, a ray of light, to cause it to break into flower. To clarify our ideas let us consider a single case, one which sums up everything. We continue from force of habit to think of the Parousia, whereby the Kingdom of God is to be consummated on Earth, as an event of a purely catastrophic nature—that is to say, liable to come about at any moment in history, irrespective of any definite state of Mankind. This is one way of looking at the matter. But why should we not assume, in accordance with the latest scientific view of Mankind in an actual state of anthropogenesis,[3] that the parousiac spark can, of physical and organic necessity, only be kindled between Heaven and a Mankind which has biologically reached a certain critical evolutionary point of collective maturity?

For my own part I can see no reason at all, theological or traditional, why this "revised" approach should give rise to any serious difficulty. And it seems to me certain, on the other hand, that by the very fact of making this simple readjustment in our "eschatological" vision we shall have performed a psychic operation having incalculable consequences. For if truly, in order that the Kingdom of God may come (in order that the Pleroma may close in upon its fullness), it is necessary, as an essential physical condition,[4] that the human Earth should already have attained the natural completion of its evolutionary growth, then it must mean that the ultra-human perfection which neo-humanism envisages for

[3] And, it may be added, in perfect analogy with the mystery of the first Christmas which (as everyone agrees) could only have happened between Heaven and an Earth which was *prepared*, socially, politically and psychologically, to receive Jesus.
[4] But not, of course, sufficient in itself!

Evolution will coincide in concrete terms with the crowning of the Incarnation awaited by all Christians. The two vectors, or components as they are better called, veer and draw together until they give a possible resultant. The supernaturalizing Christian Above is incorporated (not immersed) in the human Ahead! And at the same time Faith in God, in the very degree in which it assimilates and sublimates within its own spirit the spirit of Faith in the World, regains all its power to attract and convert!

I said at the beginning of this paper that the human world of today has not grown cold, but that it is ardently searching for a God proportionate to the new dimensions of a Universe whose appearance has completely revolutionized the scale of our faculty of worship. And it is because the total Unity of which we dream still seems to beckon in two different directions, toward the zenith and toward the horizon, that we see the dramatic growth of a whole race of the "spiritually stateless"—human beings torn between a Marxism whose depersonalizing effect revolts them and a Christianity so lukewarm in human terms that it sickens them.

But let there be revealed to us the possibility of believing *at the same time and wholly* in God *and* the World, the one through the other;[5] let this belief burst forth, as it is ineluctably in process of doing under the pressure of these seemingly opposed forces, and then, we may be sure of it, a great flame will illumine all things: for a Faith will have been born (or reborn) containing and embracing all others—and, inevitably, it is the strongest Faith which sooner or later must possess the Earth.

UNPUBLISHED. LES MOULINS (PUY-DE-DÔME), SEPTEMBER 8, 1949.

[5] In a Christ no longer seen only as the Savior of individual souls, but (precisely because He is the Redeemer in the fullest sense) as the ultimate Mover of anthropogenesis. (see diagram, p. 269).

DIAGRAM ILLUSTRATING THE CONFLICT BETWEEN THE

TWO KINDS OF FAITH IN THE HEART OF MODERN MAN

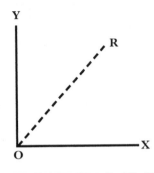

OY: Christian Faith in a personal transcendent, aspiring upward toward the Above.

OX: Human Faith in an ultra-human, driving forward toward the Ahead.

OR: Christian Faith, "rectified" or "made explicit," reconciling the two: salvation (outlet) at once Above and Ahead in a Christ who is both Savior and Mover, not only of individual men but of anthropogenesis as a whole.

Let it be noted that by its construction OR is not a half-measure, a compromise between Heaven and Earth, but a resultant combining and fortifying, each through the other, two forms of detachment—that is to say, of "sacrifice to that which is greater than self."

CHAPTER 19

ON THE PROBABLE
EXISTENCE AHEAD OF
US OF AN "ULTRA-HUMAN"

(Reflections of a Biologist)

1. *Physicobiological Definition of the Human: A Specific Superstate of Living Matter*

AT A FIRST approximation, if to begin with we try to observe it from a purely experimental standpoint, the *Human* amounts to no more than a particular fragment of matter brought locally to a state of extreme complexity or (which seems to be only another aspect of the same phenomenon) extreme "corpuscularity": the effect of its elaboration being to bring about the positive predominance, on a reflective level, of the purposeful operation of individual centers of action over the workings of hazard and large numbers.

On a reflective level: that is what it is important to understand.

From the beginning and throughout the history of the vitalization of matter we see the progressive growth of what is in fact psychism within more and more complex and interiorized organic systems (living supermolecules). Millions of years before the birth of Man, the animal felt, discovered and *knew*; but its consciousness remained simple and direct. Only Man upon this earth, completing the circle of knowledge in the depths of himself, *knows that he knows*—with the multiplicity of consequences that we all experience, without having fully assessed their stupendous biological significance: prevision of the future, construction of ordered systems, power of planned invention, regulation (and rebounding) of the evolutionary process, etc. . . . We often hear talk of the "intellectual fringe of instinct" and "animal intelligence." Such expressions, to the extent that the word "intelligence" is interpreted in its full sense of reflective psychism, are scientifically false and dangerous inasmuch as they conceal or invalidate the formidable event represented by the "punctiform" infolding of a psychic core upon itself—that is to say, the progress of Consciousness from the first to the second stage of its powers. It is, of course, perfectly legitimate to regard all the biological stems composing the Biosphere as proceeding equally, each according to its own orientation, in the universal direction of considered thought. But what is even more certain, so that it becomes self-evident when we observe in the light of facts the revolutionary biological superiority of Thought over Instinct, is that if a given Phylum X, shall we say, preceding the anthropoids, had suceeded in passing the barrier separating reflective consciousness from direct consciousness, Man would never have come into existence: instead of him, Phylum X would have woven and constituted the Noosphere.

In short, sharply sundered by a critical surface of transformation from the layers of organized matter in which it exists, the Human does indeed represent, at the heart (or summit) of Life, a core of "hyperpsychized" cosmic substance, perfectly defined and instantly recognizable by its growing, pervasive power of reasoned autoevolution which, so far as we know, it is unique in possessing.

2. Growth of the Human: The Historical Phase of Expansion

THE FACT THAT a crucial discontinuity exists between the purely animated envelope of the Earth and its thinking envelope (i.e., between the Biosphere and the Noosphere), which is manifest in the fundamentally different proceedings of Life on either side of this gap between the two layers, naturally does not mean that the Human sprang into existence among the Living in an immediate state of completeness. On the contrary.

In the first place, there is a wide zone of obscurity at the level of the zoological surface of hominization, seemingly impenetrable to our most searching methods of investigating the past, in which we can discern almost no *outward* evidence of the distinction between Irreflection and Reflection, although this had already been born. The human fetus in the earliest stages of its development, when it is virtually incapable of morphological definition, is a case in point. Even if by some inconceivable chance we were to come upon their traces, can we be sure that we should be able to recognize the first thinking beings, either by their bones or the products of their activity?

And even if we confine ourselves to those increasingly well-defined zones in which Mankind (become almost worldwide by the beginning of the Quaternary era) spreads and grows as it approaches nearer to ourselves, the fact is surely undeniable—moreover, it is a phenomenon loaded with consequences—that in a Mankind in process of planetary expansion, the Human (or Reflective) behaves like a physical magnitude, not simply variable but irresistibly rising: like a vapor which becomes increasingly gaseous as it departs in a rising temperature from its point of liquefaction.

Let us note the two main phases, one anatomical, the other social, of this progressive transhominization.

a   The anatomical phase. A question that has long been debated among anthropologists is that of how far, given the osteological characteristics of a particular cranium, it is possible to deduce, not only the shape of the brain housed within the cranium, but the particular kind of psychism contained in the brain. It is increasingly clear that in the present state of our knowledge the process is beset by innumerable pitfalls. Nevertheless, in certain cases and within approximate limits, the attempt can yield useful results.

Generally speaking the brain of mammals, as hominization takes place, not only suddenly increases its average size, and exhibits a specific development in its frontal region, but also increases what might be called its "compacity"; either structurally, by the development of the areas of association included between the sensory areas, or geometrically by the global folding back of the lobes and hemispheres upon each other.

It seems logical, this being so, to distinguish osteologically, at the beginning of anthropogenesis, a "prehominian" fossil stage represented by crania markedly less curved upon themselves (markedly less "globular") than is the case with modern man: a distinction which is borne out by the very significant fact that at this prehominian stage Mankind seems to have been made up of a more or less divergent sheaf of ethnic shoots ("subphyla"?), very much more independent of one another than any racial groups have since been. These indications suggest that the Reflective element, although already discernible in that remote period, had not yet attained the degree of perfection in its functioning that it possesses today.

In fact, it is not until we reach the artist populations of the Upper Pleistocene period—the natural scientists' *Homo sapiens*—that we really come, in a cerebral and phyletic sense, to the Human in full course of organic consolidation upon itself.

b *The Social Phase.* I say deliberately "in full course" and not "in a full state of completion": because (and this is what we must realize) it is at this point, in order to ensure the continuance of the process of hominization, that the *social* element subtly enters to take the place of the "anatomical," whose advance is at least temporarily arrested.

A great many of our contemporaries, perhaps the majority, still regard the technico-cultural knitting together of human society as a sort of para-biological epiphenomenon very inferior in organic value to other combinations achieved on the molecular or cellular scale by the forces of Life. But in terms of sound science this tendency to minimize its importance is wholly unwarrantable. For if the distinguishing characteristic of authentically "vital" arrangements of matter is that their "psychic temperature" rises proportionately to their degree of complexity, how can we withhold the status of organisms (in the fullest sense) from the groupings, so strongly "psychogenic" in their nature, which are effected within the human mass by the action of socialization?

By this interpretation, it seems to me, nothing is more wrong than to treat the Human as though it has been biologically stationary since the ending of the Ice Age. It may be that to macroscopic observation nothing has changed during this period in the generalized arrangement of the cerebral neurons. But on the other hand, what an extraordinary and irreversible increase of collective consciousness is manifest in the appearance, association and opposition of techniques, visions, passions and ideas! What an intensification of reflective life!

Indeed the grand problem with which Man confronts the biologist is not that of deciding whether anthropogenesis has effectively continued its physico-psychic progress during the past thirty thousand years, since the answer to this is clearly evident. It is far more a matter of determining whether, having reached the physical boundaries of its geographical expansion, Mankind may not be in process of leveling out, its vitality undermined by the very excess of its dimensions.

And here a great surprise awaits us.

3. Growth of the Human: The Modern Phase of Supercompression

FOR US TO regard present-day Mankind as socially complete, it would be necessary for it, having achieved the limit of planetary expansion as it has now done, to show an appreciable waning of its power of numerical increase. But what we are witnessing is the exact opposite. The statistics for a century past show no gradual stabilization but an immense rise in the earth's population. It is as though, being geographically compressed, Mankind were rebounding numerically upon itself; with the result that we can now clearly see the tightly meshed mechanism which from the first, although more or less obscurely, has governed its development.

At the beginning of the process we have increasingly severe demographic pressure, forcing the human mass to adapt itself as best it can to the confined surface of the earth: an inescapable, mathematical compression, of necessity entailing a concerted effort by individuals to find means of organizing their communal lives by arranging the world around them. So we have inventive effort and in the end the growth of Reflection (i.e., that which is Human) within the Noosphere: a growth which is again transformed (to the degree in which the increasingly reflective and hominized individual acquires a larger radius of influence and greater powers of action) into a further increase of planetary compression. So it goes on: ever-tightening compression compelling ever more Reflection . . .

From the first beginnings of History, let me repeat, this principle of the compressive generation of consciousness has been ceaselessly at work in the human mass. But from the moment—we have just reached it!—when the compression of populations in the teeming continents gains a decided ascendancy over their movement of expansion upon the earth's surface, the process is naturally speeded up to a staggering extent. We are today witnessing a truly explosive growth of technology and research, bringing an increasing mastery, both theoretical and practical, of the secrets and sources of cosmic energy at every level and in every form; and, correlative with this, the rapid heightening of what I have called the psychic temperature of the earth. A single glance at the overall picture of surface chaos is enough to assure us that this is so. We see a human tide bearing us upward with all the force of a contracting star; it is not slack water, as we might have thought, but the very crisis of the rising tide in full flood: the ineluctable growth on our horizon of a true "Ultra-Human."

4. The Face of the Ultra-Human

CLEARLY, IN THE light of what I have said, we have no grounds for expecting any relaxation, still less any end, of the process of compressive socialization which has now begun; and this being so it is fruitless to seek to escape the whirlwind that is closing in on us. What is of extreme importance, on the other hand, is that we should know what course to steer, and how we must spiritually conduct ourselves if we are to ensure that the totalitarian embrace which enfolds us will have the effect, not of dehumanizing us through mechanization, but (as seems possible) of superhumanizing us by the intensification of our powers of understanding and love.

The study of this vital question will enable us to define both the physical conditions necessary to the realization of the Ultra-Human and (to some extent!) its probable final aspect.

It may be said that for a long time, under pressure of the external forces engaged in concentrating it, the Human developed in a fashion that was mainly automatic—spurred on principally, in Bergson's expression, by a *vis a tergo*, a "push from behind." But when intelligence, which originally, as has been well said, was simply a means of survival, became gradually elevated to the function and dignity of a "reason for living," it was inevitable that, with the accentuation of the forces of liberty, a profound modification should become discernible in the working of anthropogenesis, and one of which we are only now beginning to experience the full effects. No doubt it is true that certain inward necessities, persisting in the most spiritualized recesses of our being, inexorably compel us to continue our forward progress. What power on earth has ever succeeded in arresting the growth of an idea or a passion, once they have taken shape? But the fact remains that, as Reflection increases, there is added and allied to this basic determinism the possibility of Man's withdrawal or rejection of whatever does not appear to satisfy his heart or his reason. Which is to say that, given a sufficient degree of hominization, the "planetary sequence" generating the Human can only continue to operate in an atmosphere of *consent*—meaning, finally, under the impulsion of some desire. So that in line with, and gradually replacing, the thrust from below, we see the appearance of a force of attraction coming from above which shows itself to be organically indispensable for the continuance of the sequence, indispensable for the maintenance of the evolutionary impetus, and also indispensable for the creation of an atmosphere enveloping Mankind in process of totalization, of psychic warmth and kindness without which Man's economic-technological grip upon the World can only crush souls together, without causing them to fuse and unite . . . The "pull" after the "push," as the English would say.

But whence may we expect it to come, this mysterious and indispensable force of attraction, exerting its radiance upon our minds and hearts?

In broad terms it may be affirmed that the Human, by reason of its structure, having become aware of its uncompleted state, cannot submit without extreme reluctance, still less give itself with passion, to the movement that is bearing it along unless there be some kind of discernible and definitive consummation to be looked for at the end, if only as a limit. Above all it rejects dispersal and dissolution and the circle from which there is no escape. The only air which Reflection can breathe must, of vital necessity, be that of a psychically and physically *convergent* Universe. There must be some peak, some revelation, some vivifying transformation at the end of the journey. Ultimately, and even under the urge and spur of material necessity, only a prospect, a hope of this kind is capable of sustaining our forward progress to the end.

But how exactly are we to imagine it, how conceive it, this awaited peak, this culmination of anthropogenesis, without which we shall refuse, and ever more stubbornly, to move at all?

Here we are confronted by two partly divergent and opposed answers: not merely theoretical and abstract solutions, but eventualities that have been slowly maturing in the experience of Mankind throughout the ages, and have now been abruptly brought into the daylight of our consciousness by the sudden emergence of the totalizing forces to which we are compelled to adapt ourselves.

According to the first answer (the "collectivist solution") it will suffice to ensure the biological success of our evolution if the Human organizes itself gradually on a global scale in a sort of closed circuit, within which each thinking element, intellectually and affectively connected with every other, will attain to a maximum of individual mastery by *participating* in a certain ultimate clarity of vision and extreme warmth of sympathy proper to the system as a whole. A higher state of consciousness diffused through the ultra-technified, ultra-socialized, ultra-cerebralized layers of the human mass, but without the emergence (neither necessary nor conceivable) at any point in the system of a universal, defined and autonomous Center of Reflection: this, by the first hypothesis, is all we are entitled to look for or desire as the eventual highest end of hominization.

According to the second answer on the other hand (the "personalist solution") a Center about which everything will be grouped, a keystone of the vault at the summit of the human edifice, is precisely what we must look for and postulate with all our strength, in order that nothing may crumble. For according to the supporters of this second theory, if a real power of love does not indeed arise at the heart of Evolution, stronger than all individual egotisms and passions, how can the Noosphere ever be stabilized?

And if a nucleus of ultra-consistence does not emerge at the heart of the cosmic movement, by its presence ensuring the ultimate preservation of all the incommunicable sum of Reflection sublimated through the ages by anthropogenesis, how shall we be persuaded (even under the external pressure of planetary shrinkage) to embark upon a journey leading to total Death? Indeed, to fuse together the human multitude (even taken in its present state of supercompression) without crushing it, it seems essential that there should be a field of attraction at once powerful and irreversible, and such as cannot emanate collectively from a simple nebula of reflecting atoms, but which requires as its source a self-subsisting, strongly personalized star.

This, at least by implication, is the sense of Christian argument and feeling during two thousand years. Moreover I am convinced that it is a belief that the urgency of events will increasingly compel biologists and psychologists to adopt. So that the greatest event in the history of the Earth, now taking place, may indeed be the gradual discovery, by those with eyes to see, not merely of Some Thing but of *Some One* at the peak created by the convergence of the evolving Universe upon itself.

MANY PEOPLE, OF course, will hesitate to accompany me so far in my inferences and predictions. But to keep within the bounds of what is indisputable, it appears to me that from the facts I have set forth above two conclusions emerge which must be accepted by anyone who does not refuse to see what is happening in the world today.

The first is that the Human (or, what comes to the same thing, the Reflective) not only genuinely represents, in the physical sense, a definite segregation of the "stuff of the cosmos" raised to a higher (and ever-rising) state of complexity and consciousness; but also that this separate Human element cannot achieve its final equilibrium except by coiling and concentrating, through both compulsion and attraction, on a planetary scale upon itself, until it becomes a natural unity, organically and psychically indivisible.

The second conclusion is that, in terms of this ultimate state of organization and interiorization, our present condition is still so immature that Mankind in its existing form (and although there is nothing more "adult" of its kind in the Universe with which, thus far, we can compare it) cannot be scientifically regarded as anything more than an organism which has not yet emerged from the embryonic stage.

So that in any event, whether personally centered or acentered in its eventual form (we may leave that question open) a vast realm of the Ultra-Human lies ahead of us: a realm in which we shall not be able to survive, or superlive, except by developing and embracing on earth, to the utmost extent, all the powers of common vision and unanimization that are available to us.

UNPUBLISHED. PARIS, JANUARY 6, 1950.

CHAPTER 20

HOW MAY WE CONCEIVE AND HOPE THAT HUMAN UNANIMIZATION WILL BE REALIZED ON EARTH?

HOW DEPRESSING IS the spectacle of the scattered human mass! A turbulent ant hill of separate elements whose most evident characteristic, excepting certain limited cases of deep affinity (married couples, families, the team, the mother country) seems to be one of mutual repulsion, whether between individuals or groups. Yet we nurse in the depths of our minds and hearts the conviction that it could be otherwise, that such chaos and disorder are "against nature" inasmuch as they prevent the realization, or delay the coming, of a state of affairs which would multiply as though to infinity our human powers of thought, feeling and action.

Is the situation really desperate, or are there reasons for believing in view of certain definite indications, despite appearances to the contrary, that Mankind as a whole is not only capable of unanimity but is actually in process of becoming truly unanimized? Do there exist, in other words, certain planetary energies which, as a whole, overcoming the forces of repulsion that seem to be incurably opposed to universal human harmony, are tending inexorably to bring together and organize upon itself (unbelievable though this may seem) the terrifying multitude of thousands of millions of thinking consciousnesses which forms the "reflective layer" of the earth?

My object here is to show that such energies do exist.

They are of two kinds: forces of compression, which by external and internal determinisms bring about a first stage of *enforced* unification; and subsequently forces of attraction, which through the action of internal affinity effect a genuine unanimization by *free consent.*

Let us look in turn at these two processes which so universally pervade the human atmosphere that, like light and air, we often tend to ignore them, although they envelop us so closely that no act of ours can escape them.

1. Unification by Force or Compression: The Geographical and the Mental Curvatures

a *The Geographical Curvature.* Biologically speaking the human zoological group is developing on a closed surface. More exactly, since although the world population has already virtually filled the continents to saturation-point it shows no sign of leveling out but continues to increase at an ever-growing rate, the group behaves as though it were developing in a world that is continually shrinking, so that it becomes ever more tightly compressed upon itself.

The first and obvious effect of this tremendous ethnic compression is to bring bodies relentlessly together. But the growing density of human stuff, however material its origin, is also having a profound effect on human souls. In order to respond vitally and adapt itself comfortably to the increasing pressure, in order to survive and enjoy well-being the multitude of thinking beings reacts naturally *by arranging itself* as well as possible, economically and technologically, upon itself. This automatically compels it to be constantly *inventing* new systems of mechanical equipment and social organization. In other words it is forced to reflect; and this causes it to reflect a little more upon itself, that is to say to develop further in itself those qualities which are specifically and in a higher sense human.

It is a profoundly instructive and mysterious phenomenon. The human mass is spiritually warmed and illumined by the iron grip of planetary compression; and the warming, whereby the rays of individual interaction expand, induces a further increase, in a kind of recoil, of the compression which was its cause . . . and so on, in a chain-reaction of increasing rapidity.

Out of this there arises first an irresistible grouping principle which, in its impact on the intelligence, almost automatically overrules the egoistical and mutually repulsive tendencies of the human individual.

But that is not all: for to this first geographical compression there is rapidly added a tightening effect, due this time to the emergence and influence of a curvature which is not geometric but *mental,* and which I must now explain.

b The Mental Curvature. In the "humanizing" chain of events which we have disclosed and described, the mind, which at first seemed to be no more than a "device" for confronting and resisting planetary compression, is very soon automatically transformed into a "reason" of existence. We think first in order to survive, and then we live in order to think: such is the fundamental law of anthropogenesis which emerges. But Thought, once it is let loose, displays an extraordinary power of self-protraction and extension, as though it were an independent organism which, being once born, cannot be restrained from growing and propagating itself and absorbing everything into its network. All history bears witness to the fact that nothing has ever been able to prevent an idea from growing and spreading and finally becoming universal. The reflective, psychic environment which surrounds us is so constituted that we cannot remain in it without moving forward; and we cannot advance except by drawing closer and rubbing shoulders with one another. It is as though all our individual strivings after more truth soared upward into a mental "cupola" whose closed walls inexorably compel our minds to mingle!

An enforced coalescence of all Thought in the sum total of itself . . .

The increasingly apparent growth, overriding the monstrous and chaotic human dispersal which so distresses us, of this force of auto-unification emerging from the psychic energies released by the technico-social concretion of the earth: this surely is a guarantee that, within our universe, the impulse of totalization must eventually triumph over the impulses of dispersal.

Assuredly. But on one condition. Under the influence of economic forces and the intellectual reasons invoked to break down the barriers behind which our egotism shelters, there must emerge, since this alone can be completely unanimizing, the sense of a single, fundamental aspiration.

2. Free Unification Through Attraction: A Point of Universal Convergence on the Horizon

DESPITE THE COMPULSIONS, both geographical and psychic, which oblige men to live and think in an ever closer community, they do not necessarily love each other the more on that account—

far from it. The two greatest scientists in the world, being preoccupied with the same problem, may nonetheless detest each other. This is a strange and sad fact of which we are all aware, and because of this separation of head and heart we are bound to conclude that, however social necessity and logic may impel it from behind, the human mass will only become thoroughly unified under the influence of some form of *affective* energy which will place the human particles in the fortunate position of being unable to love and fulfill themselves individually except by contributing in some degree to the love and fulfillment of all: to the extent, that is to say, that all are equal and integral parts of a single universe that is vitally converging. A "pull," in other words, must be born of the "push."[1] But amid the politico-social crisis which now besets us, have we valid, objective reasons for believing in the possibility of this happy state of affairs, even to the point of discerning its first indications?

I believe we have, on the following grounds.

If we look for the principal outcome, "Result No. 1," of the ineluctable scientific unification of our intellects during the past century, we must quickly perceive that the gain consists far less in our securing control of any particular source of natural energy than in the general awakening of our consciousness to the vast and extreme organicity of the universe as a whole, considered in terms of its internal forces of development. We see more clearly with every increase in our knowledge that we are, all of us, participants in a process (Cosmogenesis culminating in Anthropogenesis) upon which our ultimate fulfillment—one might even say, our beatification—obscurely depends. And whence can it arise, this accumulation of evidence that the extreme point of each of us (our

[1] Teilhard uses the English words.

ultra-ego, it might be termed) coincides with some common fulfillment of the evolutionary process, a common super-ego, except out
of the principle of attraction which we have postulated and invoked above as being necessary to make the rebellious nuclei of
our individual personalities cohere from within, to instil unanimity
even in their hearts?

Thus, superimposed on the twofold tightening action of what
I have called the geometrical and mental curvatures of the human earth—superimposed and *emanating* from them—we have a
third and final unifying influence brought to bear in regulating
the movements of the Noosphere, that of a destiny that is
supremely attractive, the same for all at the same time. A total
community of desire, which makes of it a third force as planetary
in its dimensions as the other two, but operating, no matter how
irresistibly, in the manner of a seduction—that is to say, by free
consent.

It would be premature to assert that this new force as yet plays
any very explicit part in the course of political or social events
around us. Yet may we not claim, observing the precipitate growth
and succession of democracies and totalitarian regimes during the
past hundred and fifty years in the history of the world, that it is
the *Sense of the Species*, which for a time seemed to have vanished
from the depths of our hearts, dispelled in some sort by the growth
of Reflection, that is now gradually resuming its place and re-
asserting its rights over all narrow individualism? The Sense of the
Species interpreted in the new, grand human manner: not, as formerly, a shoot which merely seeks to prolong itself until it bears its
fruit, but the fruit itself, gathering and growing upon itself in the
expectation of eventual ripeness.

But if the hope of this maturing of the Species, and the belief
in its coming, are to illumine and truly unanimize our hearts, we

must endow it with certain positive attributes. It is here that opinions are divided.

Those who think on Marxist lines believe that all that is necessary to inspire and polarize the human molecules is that they should look forward to an eventual state of *collective* reflection and sympathy, at the culmination of anthropogenesis, from which all will benefit through *participation*: as it were, a vault of mutually reinforced thoughts, a closed circuit of attachments in which the individual will achieve intellectual and affective fulfillment to the extent that he is one with the whole system.

But in the Christian view only the eventual appearance, at the summit and in the heart of the unified world, of an autonomous center of congregation is structurally and functionally capable of inspiring, preserving and fully releasing, within a human mass still spiritually dispersed, the looked-for forces of unanimization. According to the supporters of this hypothesis only a veritable *super-love*, the attractive power of a veritable "super-being," can of psychological necessity dominate, possess and synthesize the host of other earthly loves. Failing such a center of universal convergence, not metaphorical or potential but *real*, there can be no true coherence among totalized Mankind, and therefore no true consistence. A world culminating in the Impersonal can bring us neither the warmth of attraction nor the hope of irreversibility (immortality) without which individual egotism will always have the last word. A veritable *Ego* at the summit of the world is needed for the consummation, without confounding them, of all the elemental *egos* of Earth . . . I have talked of the "Christian view," but this idea is gaining ground in other circles. Was it not Camus who wrote in *Sisyphe*, "If Man found that the Universe could love he would be reconciled"? And did not Wells, through his exponent the humanitarian biologist Steele in *The Anatomy of Frustration*, express his need to find, above and beyond humanity, a "universal lover"?

Let me recapitulate and conclude.

Essentially, in the twofold irresistible embrace of a planet that is visibly shrinking, and Thought that is more and more rapidly coiling in upon itself, the dust of human units finds itself subjected to a tremendous pressure of coalescence, far stronger than the individual or national repulsions that so alarm us. But despite the closing of this vise nothing seems finally capable of guiding us into the natural sphere of our interhuman affinities except the emergence of a powerful field of internal attraction, in which we shall find ourselves caught *from within*. The rebirth of the Sense of Species, rendered virtually inevitable by the phase of compressive and totalizing socialization which we have now entered, affords a first indication of the existence of such a field of unanimization and a clue to its nature.

Nevertheless, however efficacious this newly born faith of Man in the ultra-human may prove to be, it seems that Man's urge toward *Some Thing* ahead of him cannot achieve its full fruition except by combining with another and still more fundamental aspiration—one from above, urging him toward *Some One*.

UNPUBLISHED. PARIS, JANUARY 18, 1950.

CHAPTER 21

FROM THE PRE-HUMAN TO THE ULTRA-HUMAN: THE PHASES OF A LIVING PLANET

ASTRONOMY IS BEGINNING to detect and classify in the heavens a life of the stars, red, blue and white, giant, middle-sized and dwarf; each type, in its dimensions, particular radiations and brilliance, being subject to a given evolutionary cycle.

It is a matter of great interest; but have we sometimes thought how much more interesting and moving it would be if we could observe or at least reconstruct the history, not of the glowing suns in the heart of galaxies but of the mysterious living planets? Celestial bodies such as these (they undoubtedly exist as we shall see) give out no perceptible radiation, or none that our present instruments can detect.[1] We know nothing as yet of their number, their distribution or their history. Our study of them, in short, is restricted to a single specimen, that of our own Earth, which is apparently far from having attained its full development.

[1] But is it inconceivable that there should some day be spectroscopes sensitive to some form of vital radiation?

It is an unfavorable situation, but capable nevertheless of being put to use, since by means of the remarkable phenomena of sedimentation and fossilization we can trace the biological past of this planet over a period of nearly a thousand million years.

Using it as a representative example, though still unique in our experience and probably "immature," let us seek to sketch on scientific lines the probable evolutionary curve of any living planet; a problem in which affective reasoning is singularly mingled with speculation, since what we are looking for and seeking to extrapolate is nothing less than our own destiny.

More than 600 million years ago the earth, like a nova of a singular kind, began to glow dimly with life. Under the influence of solar radiation the sensitive film of its youthful waters became charged in places with asymmetrical and multiplying proteins. We do not know what caused this phenomenon, whether it was the outcome of some sudden convulsion or of a long process of ripening. What we do know is that this did indeed happen, and moreover that of statistical necessity it could not have failed to happen, given the physicochemical conditions prevailing on the planet that bears us. However improbable in a mechanistic sense the elaborate organic structure created by life may appear, it seems increasingly evident that the cosmic substance is drawn toward these states of extreme arrangement by a particular kind of attraction which compels it, by the play of large numbers in which it is involved, to miss no opportunity of becoming more complex and thus achieving a higher degree of freedom.

So we may assume that sporadically, in the course of time, numerous centers of indeterminacy and consciousness can and must have appeared in sidereal space, of which our own Earth is one. Although Life by its structure seems in certain ways to be highly exceptional, everything suggests that its pressure is exerted throughout the universe. And everything suggests that, wherever cosmic hazard has enabled it to hatch out and establish itself, it cannot thereafter cease to become intensified to the utmost, in accordance with an automatic process which may be analyzed as follows:

First, increase. Even in its lowest forms living matter, by its physicochemical nature, possesses the extraordinary power of reproducing itself indefinitely in a geometrical progression. For this reason however minute and scattered the first patches of vitalized proteins may have been, they could do no other than spread rapidly until they covered the entire surface of the planet; and their expansion within a closed circle, after its initial unrestricted stage, produced an increasing degree of compression. A gas under mounting mechanical pressure as a rule changes its state. In the same way a multitude of individuals, a living mass, being subjected to pressure within an enclosed space, and to increasing biological interpenetration, reacts by organizing itself upon itself: that is to say, by seeking through selective experiment for the individual or collective arrangements which best suit it, such arrangements being, in the event, those in which the degree of complexity is highest and therefore the state of indeterminacy the most advanced.

Many biologists, intent upon scientific objectivity are reluctant to see in the historical development of terrestrial life anything more than an unlimited proliferation of forms, all on the same level. A steady increase of living creatures and living combinations, they agree; but, despite this, not more life. What reason have we for supposing that a mammal is *more* than a polypary?

Far more suggestive and convincing than this "flat" vision of the biological world is the three-dimensional concept of a heavenly body on which, through the effect of planetary compression, the state of complexity (or, which amounts to the same thing, the "psychic" temperature of the biosphere) is continually rising. This explains the supersession in successive stages of arthropods by vertebrates, of pisciforms by tetrapods, and finally, within the tetrapod group, the progressive predominance of the mammals, gradually forming their own primate strain, with the growth, globally irreversible and constantly accelerated along certain favored lines, of "cerebration" from the beginning of life up to the present time. The quantity and quality of cephalized nervous substance on earth have indeed never been as great as they are today. This "orthogenetic" view of animal evolution is gradually becoming common ground among scientists; but it only achieves full validity, in terms of my argument, to the extent that it implies a continuous psychic "chain" going back to the beginning of life.

Looking back over the immense extent of geological time we can see that the separate links in the chain have undergone no essential change. It would seem that the principal factor making for progress is still the operation of forces of natural selection, choosing from outside the most successful and adaptable products of a process of expansion that is disorderly in itself. Where, during the course of time, an important transformation seems to have taken place is at the level of the latest link in the chain, that of the "acquirement of consciousness." For it is inevitable, by very reason of the selective growth of psychism in the biosphere, that each new higher element engendered by evolution must, to the extent that it is more conscious, have a wider field of action. The mere fact of its "ultra-cerebration" causes it to take up more room. So that the compression of living matter, due in the first place simply to physical increase, is gradually heightened by its internal psychic expansion. The chain coils in upon itself and the intensity of the phenomenon tends to rise almost vertically Or to adopt another image one might say that the "psychic tint" of the earth, studied at a great distance by some celestial observer, for two combined reasons would be seen, in the course of eons of geological time, to become gradually heightened in intensity until it reaches the peculiarly moving moment of climax when, in a spread of more active radiation covering Africa and southern Asia, a series of sparks begins to glow, foreshadowing the incandescence which is "hominization."

Closely related though he is to the other major primates, among which he is biologically only one of a family, Man is psychically distinguished from all other animals by the entirely new fact that he not only knows, but knows that he knows. In him, for the first time on earth, consciousness has coiled back upon itself to become thought. To an observer unaware of what it signifies, the event might at first seem to have little importance; but in fact it represents the complete resurgence of terrestrial life upon itself. In reflecting psychically upon itself Life positively made a new start. In a second turn of the spiral, tighter than the first, it embarked for a second time upon its cycle of multiplication, compression and interiorization.

This is how the thinking layer of the earth as we know it today, the Noosphere, came rapidly into being, proceeding from certain centers of reflection which apparently emerged, at the threshold of the Pleistocene period, somewhere in the tropical or subtropical zones of the Ancient World[2]: a planetary neo-envelope, essentially linked with the biosphere in which it has its root, yet distinguished from it by an autonomous circulatory, nervous and, finally, cerebral system. The Noosphere: a new stage for a renewed Life.

One may say that until the coming of Man it was natural selection that set the course of morphogenesis and cerebration, but

---

[2] i.e., in the place where, during the Upper Tertiary era, the group of the great anthropoids was first established and subsequently spread.

that after Man it is the power of invention that begins to grasp the evolutionary reins. A wholly inward change, having no direct effect on anatomy; but a change, as we now know, entailing two decisive consequences for the future. The first is an unlimited increase in the aura of influence radiating from every living being; the second, even more radical, the prospect afforded to a growing number of individuals of being joined together and ever more closely unanimized in the inextinguishable fire of research pursued in common.

From the Quaternary era onward Life has continued to superdevelop itself, through Man, in the second degree. But although this phenomenon is several hundred thousand years old, there are growing indications that the process, far from slowing down, is now entering upon a particularly accelerated and critical phase of its development.

So far as we are able to follow its historical progress, the grouping and organization of the human mass has in the past been broadly governed far more by the principle of expansion than by that of compression. Diverse civilizations were able to grow and rub shoulders on a sparsely inhabited planet without encountering any major difficulty. But now, following the dramatic growth of industry, communications and populations in the course of a single century, we can discern the outline of an astonishing event. The hitherto scattered fragments of humanity, being at length brought into close contact, are beginning to interpenetrate to the point of reacting economically and psychically upon each other; with the result, given the fundamental relationship between biological compression and the heightening of consciousness, of an irresistible rise within us and around us of the level of Reflection. Under the influence of the forces compressing it within a closed vessel, human substance is beginning to "planetize" itself, that is to say, to be interiorized and animated globally upon itself.

We may have supposed that the human species, being matured, has reached the limit of its development. Now we see that it is still in an *embryonic state*. Science can discern, in the hundreds of thousands (or more probably millions) of years[3] lying ahead of the Mankind we know, a deep if still obscure fringe of the "ultra-human."

If this is so, and assuming that no sidereal accident interferes with the course of events, how is the adventure likely to conclude? Can we look forward to nothing but a state of *senescence* at the end of the planetary cycle of hominization or, on the contrary, will it be a *paroxysm* of the Noosphere?

The senescence theory finds immediate and natural support in the fact of our individual ends. Since each separate thinking element of the earth is destined to wither and die, why should the sum total of them all, Mankind, be exempt from a similar fate? This is the first thought that occurs to us: but is it sound? Is it certain that we can extrapolate the general evolutionary curve of the species (phylogenesis) on the lines of the evolutionary pattern of the individual (ontogenesis) without making any correction? Nothing proves that we can, and there is a powerful argument against it. For although certain principles of wear and disintegration, which apparently cannot be prevented from growing more pronounced with age, seem to be inherent in the structure of our individual bodies, there is no indication of any similar factor in the global evolution of a living mass as large as the Noosphere, where the overriding evolutionary law seems to be that of statistical necessity, it must simply converge upon itself.

The more deeply we study this distinction the more probable does it seem that the human multitude is moving as time passes not

[3] Since Mankind's behavior on the "tree of Life" is rather that of a flowering than of an ordinary shoot, it is possible that the estimate of several million years, based on the average longevity of animal forms, should be materially reduced to allow for the acceleration due to the totalization of the Noosphere.

toward any slackening but rather toward a superstate of psychic tension. Which means that it is not any sluggishness of the spirit that lies ahead of us, but on the contrary an eventual critical point of collective reflexion. Not a gradual darkening but a sudden blaze of brilliance, an explosion in which Thought, carried to the extreme, is volatilized upon itself: such, if I had to bet on it, is how I would depict the ultimate phase of a vitalized star.

But can even this, a supreme explosion, be considered a biologically satisfactory culmination of the phenomenon of Man? It is here that we encounter the very root of the problem proposed to our scientific understanding by the existence of living planets.

In speaking of the rise of terrestrial psychic temperature I have always assumed that in the Noosphere, as in the Biosphere, the need and the will to grow both remain constant. There can be no natural selection, still less reflective invention, if the individual is not inwardly intent upon "superliving," or at least upon survival. No evolutionary mechanism can have any power over a cosmic matter if it is entirely passive, less still if it is opposed to it. But the possibility has to be faced of Mankind falling suddenly out of love with its own destiny. This disenchantment would be conceivable, and indeed inevitable, if as a result of growing reflection we came to believe that our end could only be collective death in an hermetically sealed world. Clearly in face of so appalling a discovery the psychic mechanism of evolution would come to a sudden stop, undermined and shattered in its very substance, despite all the violent tuggings of the chain of planetary in-folding.

The more one considers this eventuality (which cannot be dismissed as a myth, as certain morbid symptoms, such as Sartrian existentialism, show) the more does one tend to the view that the grand enigma presented by the phenomenon of Man is not the question of knowing how life was kindled on earth, but of understanding how it might be extinguished on earth without being continued elsewhere. Having once become reflective it cannot acquiesce in its total disappearance without biologically contradicting itself.

In consequence one is the less disposed to reject as unscientific the idea that the critical point of planetary Reflection, the fruit of socialization, far from being a mere spark in the darkness, represents our passage, by translation or dematerialization, to another sphere of the Universe: not an ending of the Ultra-Human but its accession to some sort of Trans-Human at the ultimate heart of things.

PARIS, APRIL 27, 1950. *ALMANACH DES SCIENCES,* 1951.

CHAPTER 22

THE END OF
THE SPECIES

NOT MUCH MORE than a hundred years ago,
Man learned to his astonishment that there was an
origin of animal species, a genesis in which he
himself was involved. Not only did all kinds of animals share the earth with him, but he found that
he was in some sort a part of this zoological diversity which hitherto he had regarded as being
merely his neighbors. Life was in movement, and
Mankind was the latest of its successive waves!

This astonishing pronouncement on the part
of science seemed at first to do no more than stim-
ulate the curiosity (or indignation) of theorists;
but it was soon apparent that the shock was not
purely mental, and that nineteenth-century man
had been shaken by it to his depths. Three hundred years earlier, in the time of Galileo, the end of
geocentrism had intrigued or disturbed thinking
minds without having any appreciable effect on the
mass of people. The sidereal dispute had, after all,

produced no change in the earth itself, or in its inhabitants or their relations with one another. But the concept of biological evolution inevitably led to a profound reshaping of planetary values.

To some outraged spirits, no doubt, Man appeared diminished and dethroned by transformism which made him no more than the latest arrival in the animal kingdom. But to the minds of the majority our human condition seemed finally to be exalted by the fact that we were rooted in the fauna and soil of the planet—evolving Man in the forefront of the animals.

In short, until then, Man, although he knew that the human race might continue to exist for a long time, had not suspected that it had a future. Now however, because he was a species, and species change, he could begin to look for and seek to conquer something quite new that lay ahead of him.

That is why "Darwinism," as it was then called, however naive its beginnings, came at exactly the right moment to create the cosmological atmosphere of which the great technico-social advance of the last century stood in need if it was to believe passionately in what it was doing. Rudimentary though it was, Darwinism afforded a scientific justification of faith in progress.

———————— ✳ ————————

BUT TODAY, BY a development natural to itself, the movement has come to look like a receding tide. For all his discoveries and inventions, twentieth-century man is a sad creature. How shall we account for his present dejected state except basically by the fact that, following that exalted vision of species in growth, he is now confronted by an accumulation of scientific evidence pointing to the reverse—the species doomed to extinction?

The extinction of the species . . .

Biologists do not agree about the mechanism of the continual disappearance of phyla in the course of geological time, a process almost as mysterious as that of their formation; but the reality of the phenomenon is indisputable. Either the different species, losing their powers of "speciation," survive as living fossils, which after all is a form of death; or else, and there are infinitely more of these, they simply vanish, one sort being replaced by another. Whatever the reason may be, inadaptability to a new environment, competition, a mysterious senescence, or possibly a single basic cause underlying all these reasons, the end is always the same. The days (or the millennia) of every living form are by statistical reckoning ineluctably numbered; so much so that, using the scale of time furnished by the study of certain isotopes, it is beginning to be possible to calculate in millions of years *the average life of a species.*

Man now sees that the seeds of his ultimate dissolution are at the heart of his being. The *End of the Species* is in the marrow of our bones!

Is it not this presentiment of a blank wall ahead, underlying all sorts of tensions and specific fears, which paradoxically (at the very moment when every barrier seems to be giving way before our power of understanding and mastering the world) is darkening and hardening the minds of our generation?

AS PSYCHIATRY TEACHES us, we shall gain nothing by shutting our eyes to this shadow of collective death that has appeared on our horizon. On the contrary, we must open them wider.

But how are we to exorcise the shadow?

It may be said that timidly, even furtively (it is remarkable how coy we are in referring to the matter) two methods are used by writers and teachers to reassure themselves and others in face of the ever more obsessive certainty of the eventual ending of the human species: the first is to invoke the infinity of Time and the second is to seek shelter in the depths of Space.

The Time argument is as follows. By the latest estimates of palaeontology the probable life of a phylum of average dimensions is to be reckoned in tens of millions of years. But if this is true of "ordinary" species, what duration may we not look for in the case of Man, that favored race which, by its intelligence, has succeeded in removing all danger of serious competition and even in attacking the causes of senescence at the root.

Then the Space argument. Even if we suppose that, by prolonging its existence on a scale of planetary longevity, the human species will eventually find itself with a chemically exhausted Earth beneath its feet, is not Man even now in process of developing astronautical means which will enable him to go elsewhere and continue his destiny in some other corner of the firmament?

That is what they say, and for all I know there may be people for whom this sort of reasoning does really dispel the clouds that veil the future. I can only say that for my part I find such consolations intolerable, not only because they do nothing but palliate and postpone our fears, which is bad enough, but even more because they seem to me scientifically false.

In order that the end of Mankind may be deferred *sine die* we are asked to believe in a species that will drag on and spread itself indefinitely; which means, in effect, that it would run down more and more. But is not this the precise opposite of what is happening here and now in the human world?

I have been insisting for a long time on the importance and significance of the technico-mental process which, particularly during the past hundred years, has been irresistibly causing Mankind to draw closer together and unite upon itself. From routine or prejudice the majority of anthropologists still refuse to see in this movement of totalization anything more than a superficial and temporary side effect of the organic forces of biogenesis. Any parallel that may be drawn between socialization and speciation, they maintain, is purely metaphorical. To which I would reply that, if this is so, to what undisclosed form of energy shall we scientifically attribute the irreversible and conjugated growth of Arrangement and Consciousness which historically characterizes (as it does everything else, in indisputably "biological" fields) the establishment of Mankind on Earth?

We have only to go a little further, I am convinced, and our minds, awakened at last to the existence of an added dimension, will grasp the profound identity existing between the forces of civilization and those of evolution. Man will then assume his true shape in the eyes of the naturalists—that of a species which having entered the realm of Thought, henceforth folds back its branches upon itself instead of spreading them. Man, *a species which converges*, instead of diverging like every other species on earth: so that we are bound to envisage its ending in terms of some paroxysmal state of maturation which, by its scientific probability alone must illumine for us all the darkest menaces of the future.

For if by its structure Mankind does not dissipate itself but continually concentrates upon itself; in other words, if, alone among all the living forms known to us, our zoological phylum is laboriously moving toward a *critical point of speciation*, then are not all hopes permitted to us in the matter of survival and irreversibility?

The end of a "thinking species": not disintegration and death, but a new breakthrough and a rebirth, this time outside Time and Space, through the very excess of unification and coreflexion.[1]

---

[1] Such coreflexion, as I am constantly obliged to say, in no way entailing a diminution but on the contrary an increase of the "person." Must I again repeat the truth, of universal application, that if it be properly ordered *union does not confound but differentiates?*

It goes without saying that this idea of a salvation of the Species sought, not in the direction of any temporo-spatial consolidation or expansion but by way of spiritual escape through the excess of consciousness, is not yet seriously considered by the biologists. At first sight it appears fantastic. Yet if one thinks about it long and carefully, it is remarkable how it sustains examination, grows stronger and, for two particular reasons among others, takes root in the mind.

For one thing, as I have said, it corresponds more closely than any other extrapolation to the marked (even challenging) urgency of our own time in the broad progress of the Phenomenon of Man. But in addition it seems to be more capable than any other vision of the future of stimulating and steadying our power of action by counteracting the prevailing pessimism.

This is a fact which we must face.

In the present age, what does most discredit to faith in progress (apart from our reticence and helplessness as we contemplate the "end of the Race") is the unhappy tendency still prevailing among its adepts to distort everything that is most valid and noble in our newly aroused expectation of an "ultra-human" by reducing it to some form of threadbare millennium. The believers in progress think in terms of a Golden Age, a period of euphoria and abundance; and this, they give us to understand, is all that Evolution has in store for us.[2] It is right that our hearts should fail us at the thought of so "bourgeois" a paradise.

We need to remind ourselves yet again, so as to offset this truly pagan materialism and naturalism, that although the laws of biogenesis by their nature presuppose, and in fact bring about, an improvement in human living conditions, it is not *well-being* but a

---

[2] I may cite, as an instance of this poverty of thought, the French film sheltering behind so many famous names, *La Vie Commence Demain*.

hunger for *more-being* which, of psychological necessity, can alone preserve the thinking earth from the *taedium vitae*. And this makes fully plain the importance of what I have already suggested, that it is upon its point (or superstructure) of spiritual concentration, and not on its basis (or infrastructure) of material arrangement, that the equilibrium of Mankind biologically depends.

For if, pursuing this thought, we accept the existence of a critical point of speciation at the conclusion of all technologies and civilizations, it means (with Tension maintaining its ascendancy over Rest to the end in biogenesis) that an *outlet* appears at the peak of Time, not only for our hope of escape but for our expectation of some revelation.

And this is what can best allay the conflict between light and darkness, exaltation and despair, in which, following the rebirth in us of the Sense of the Species, we are now absorbed.

NEW YORK, DECEMBER 9, 1952. *PSYCHÉ*, FEBRUARY 1953.

NOTE BY FRENCH EDITOR. Underlying this final testimony is Teilhard de Chardin's earliest mystical intimation, set forth in *Cosmic Life* as early as 1916. The following extracts from that work show the unity of fundamental Christian vision and scientific knowledge which he preserved to the end.

Cosmic Life[3]

GOD CANNOT IN any way be intermixed with or lost in the participated being which he sustains and animates and holds together,

---

[3] Printed in full in *Writings in Time of War*, pp. 14–71. Collins, London and Harper & Row, New York, 1968.

but he is at the birth, and the growth and the final term of all things . . .

The exclusive task of the world is the physical incorporation of the faithful in the Christ who is of God. This cardinal task is being carried out *with the rigor and harmony of a natural evolution.*

At the source of its developments an operation was called for, transcendent in order, to graft the person of a God onto the human cosmos, under conditions that are mysterious but physically governed . . . *Et Verbum caro factum est.* This was the Incarnation. From this first and fundamental contact between God and the human race—which means in virtue of the penetration of the Divine into our nature—a new life was born: an unlooked for magnification and "obediental" extension of our natural capabilities—grace . . . Grace is the unique sap that starts from the same trunk and rises up into the branches, it is the blood that courses through the veins under the impulse of one and the same Heart, the nervous current that is transmitted through the limbs at the dictate of one and the same Head: and that radiant Head, that mighty Heart, that fruitful Stock, must inevitably be Christ . . .

The Incarnation is a making new, a restoration, of *all* the universe's forces and powers; Christ is the Instrument, the Center, the End, of the *whole* of animate and material creation; through Him, *everything* is created, sanctified and vivified. This is the constant and general teaching of St. John and St. Paul (that most "cosmic" of sacred writers), and it has passed into the most solemn formulas of the Liturgy: and yet we repeat it, and generations to come will go on repeating it, without ever being able to grasp or appreciate its profound and mysterious significance, bound up as it is with understanding of the universe.

With the origin of all things, there began an advent of recollection and work in the course of which the forces of determinism, obediently and lovingly, lent themselves and directed themselves in the preparation of a Fruit that exceeded all hope and yet was awaited. The world's energies and substances—so harmoniously adapted and controlled that the supreme Transcendent would seem to germinate entirely from their immanence—concentrated and were purified in the stock of Jesse; from their accumulated and distilled treasures they produced the glittering gem of matter, the Pearl of the Cosmos, and the link with the incarnate personal Absolute—the Blessed Virgin Mary, Queen and Mother of all things, the true Demeter . . . and when the day of the Virgin came to pass, then the final purpose of the universe, deep-rooted and gratuitous, was suddenly made clear: since the days when the first breath of individualization passed over the expanse of the *Supreme Center here below* so that in it could be seen the ripple of the smile of the original monads, all things were moving toward the Child born of Woman.

And since Christ was born, and ceased to grow, and died, *everything has continued in motion because he has not yet attained the fullness of his form.* He has not gathered about Him the last folds of the garment of flesh and love woven for him by his faithful. *The mystical Christ has not reached the peak of his growth* . . . and it is in the continuation of this engendering that there lies the ultimate driving force behind all created activity . . . Christ is the term *of even the natural* evolution of living beings.

CONCLUSION

The End of the World

NOTE BY FRENCH EDITOR. To conclude these
writings on the Future of Man we quote the following extract from another work, *My Universe.*[1]
Summarizing in a luminous synthesis the thinker
and priest's intimations of the End of the World, it
ends with the words of St. Paul, quoted on the last
page of Teilhard de Chardin's journal, which express his supreme vision: "God all in all."[2]

. . . Forced against one another by the increase in
their numbers and the multiplication of their in-
terrelations—compressed together by the activation of a common force and the awareness of a
common distress—the men of the future will form,
in some way, but one single consciousness; and
since, once their initiation is complete they will
have gauged the strength of their associated
minds, the immensity of the universe, and the nar-

[1] Printed in *Science and Christ*, pp. 83–5. Collins, London, and
Harper & Row, New York, 1968.
[2] In Latin: *Erit in omnibus omnia Deus.* In Greek: *En pasi panta
Theos.* 1 Corinthians 15.28.

rowness of their prison, this consciousness will be truly adult and of age. May we not imagine that at that moment a truly and totally human act will be effected for the first time, in a final option—the yes or no as an answer to God, pronounced individually by beings in each one of whom the sense of human freedom and responsibility will have reached its full development?

It is by no means easy to picture to ourselves what sort of event the end of the world could be. A sidereal catastrophe would be a fitting counterpart to our individual deaths, but it would entail the end of the earth rather than that of the cosmos—and it is the cosmos that has to disappear.

The more I think about this mystery, the more it appears to me, in my dreams, as a "turning-about" of consciousness—as an eruption of interior life—as an ecstasy. There is no need to rack our brains to understand how the material vastness of the universe will ever be able to disappear. Spirit has only to be reversed, to move into a different zone, for the whole shape of the world immediately to be changed.

When the end of time is at hand, a terrifying spiritual pressure will be exerted on the confines of the real, built up by the desperate efforts of souls tense with longing to escape from the earth. This pressure will be unanimous. Scripture, however, tells us that at the same time the world will be infected by a profound schism—some trying to emerge from themselves in order to dominate the world even more completely—others, relying on the words of Christ, waiting passionately for the world to die, so that they may be absorbed with it in God.

It is then, we may be sure, that the Parousia will be realized in a creation that has been taken to the climax of its capacity for union. The single act of assimilation and synthesis that has been going on since the beginning of time will then at last be made plain, and the universal Christ will blaze out like a flash of lightning in the storm clouds of a world whose slow consecration is complete. The trumpets of the angels are but a poor symbol. It will be impelled by the most powerful organic attraction that can be conceived (the very force by which the universe holds together) that the monads will join in a headlong rush to the place irrevocably appointed for them by the total adulthood of things and the inexorable irreversibility of the whole history of the world—some, spiritualized matter, in the limitless fulfillment of an eternal communion—others, materialized spirit, in the conscious torment of an endless decomposition.

At that moment, St. Paul tells us (1 Cor. 15. 23 ff) when Christ has emptied all created forces (rejecting in them everything that is a factor of dissociation and superanimating all that is a force of unity), he will consummate universal unification by giving himself, in his complete and adult Body, with a finally satisfied capacity for union, to the embrace of the Godhead.

Thus will be constituted the organic complex of God and world—the Pleroma—the mysterious reality of which we cannot say that it is more beautiful than God by himself (since God could dispense with the world), but which we cannot, either, consider completely gratuitous, completely subsidiary, without making Creation unintelligible, the Passion of Christ meaningless, and our effort completely valueless.

Et tunc erit finis.

Like a vast tide, Being will have engulfed the shifting sands of being. Within a now tranquil ocean, each drop of which, nevertheless, will be conscious of remaining itself, the astonishing adventure of the world will have ended. The dream of every mystic, the eternal pantheist ideal, will have found its full and legitimate satisfaction. "*Erit in omnibus omnia Deus.*"

TIENTSIN, MARCH 25, 1924.

Conclusion

NOTE BY FRENCH EDITOR. Three days before his death Pierre Teilhard de Chardin wrote the following, which constitutes his supreme testimony as a thinker and a priest.

LAST PAGE OF THE JOURNAL OF PIERRE TEILHARD DE CHARDIN

Maundy Thursday.   *What I believe.*

1   St. Paul—the three verses: *En pasi panta Theos.*
     Christogenesis.

2   Cosmos = Cosmogenesis—Biogenesis—Noogenesis—

3   } *The universe is centered*—Evolutively   { Above
                                                     Ahead

The two
articles of } *Christ is its Center* { The Christian Phenom-
my Credo   enon
                                            Noogenesis = Christogene-
                                            sis (≡ Paul)

The three verses are 1 Corinthians 15. 26, 27 and 28:

The last enemy that shall be destroyed is death.

For he hath put all things under his feet. But when he saith, all things are put under him, it is manifest that he is expected, which did put all things under him.

And when all things shall be subdued unto him, then shall the Son also himself be subject unto him that put all things under him, that God may be all in all.

INDEX

Above, and Ahead, 262ff.
acquired characteristics,
    transmission, 17
action: human, environment of,
    48ff.; in individual and
    mankind, 8f.; problem of, 42
activation, coefficient of, 204f.
additivity, 16ff.; education as, 20
aesthetic powers, development of, 6
aggregation, human, stages of, 169f.
agricultural groups, 170
agriculture, discovery of, 135
albuminoids, 99; *see also* proteins
animals: "education" among, 18f.;
    limbs of, as tools, 158f.; *see also*
    consciousness
annihilation, point of, 47
anthropogenesis, 232n., 261f., 275,
    286
ants, 29, 31, 249
asceticism, 34f.
association(s), 29; differentiation in,
    44f.
astronomy, 28, 91f., 290
astrophysics, 92
atheism: growth of, 260; Marxist,
    266
atom(s): genesis of, 250; importance
    of links between, 104
atomic energy, release of, 133ff.
atomism, 57
attraction, forces of, 232f., 278ff.
Augustine, St., 9

autoevolution, 272
autonomy, human, 9

Babel, 182
baboon, 157n.
bees/bee colonies, 29, 249
behavior-patterns, transmission of, 20
being, value of, 32f., 48
Benda, Julien, 207n.
Bergson, Henri, 19, 106, 216, 251n.,
    277
Betti, 207n.
biology, 28, 176
biosphere, 151, 193n., 271 *et passim*
Blanc, A., 152n.
Boll, Marcel, 212
boredom, 139
brain: collective, 161f., 172f.; human,
    217; evolution of, 163; and
    social thought, 161ff.; and
    osteology, 273; prehominid, 273
Buddha, 41

Camus, Albert, 288
cells, 100, 108
centration, 154, 203
cephalization, 56
cerebralization/cerebration, 56,
    168, 293
chance, influence of, 197
change: entitative, in man, 6;
    morphological, slowing down
    of, 15

PIERRE TEILHARD DE CHARDIN (1881–1955) was a philosopher, paleontologist, and Jesuit priest. Born in France, educated in Jesuit schools, and ordained in 1911, he journeyed to various parts of the world on geological and paleontological expeditions and published several works on science. His renowned works, *The Phenomenon of Man* and *The Divine Milieu*, were published shortly after his death and today are regarded as classics of Catholic theology.